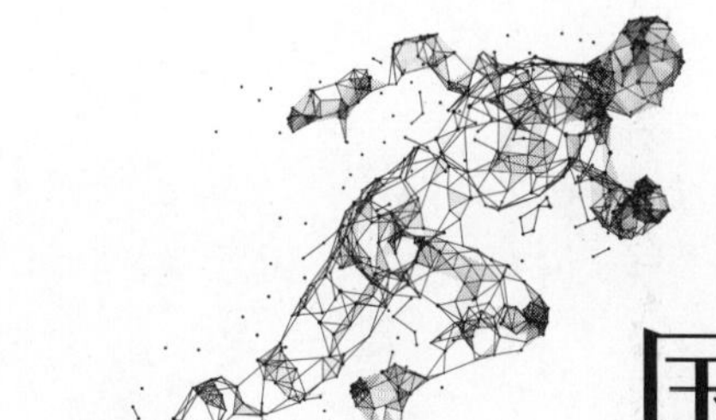

国际数学竞赛辅导

（初级）

龚朝阳　何坪阳　晏冬◎主编

清華大学出版社
北　京

内容简介

本书围绕初等数论、初等代数、初等几何和组合数学四大模块，系统梳理了国际数学竞赛中的核心知识体系和典型题型。全书采用中英文对照形式编写，有助于学生更好地适应国际数学竞赛环境。书中例题和习题丰富，讲解详尽，内容覆盖常见考点，注重知识的系统性和实用性，能够帮助学生建立完整的竞赛知识体系并且提升学生的实际解题能力。

本书不仅适用于备考 AMC 等国际数学竞赛的中学生系统学习，也可作为教师和数学竞赛教练的日常教学与集训参考。

图书在版编目（CIP）数据

国际数学竞赛辅导：初级：汉文、英文 / 龚朝阳，何坪阳，晏冬主编 . 北京：清华大学出版社，2025. 7. -- ISBN 978-7-302-69947-7

Ⅰ. O1-44

中国国家版本馆 CIP 数据核字第 2025N816K7 号

责任编辑： 陈　明
封面设计： 傅瑞学
责任校对： 赵琳爽
责任印制： 宋　林

出版发行： 清华大学出版社
网　　址： https://www.tup.com.cn，https://www.wqxuetang.com
地　　址： 北京清华大学学研大厦 A 座　　**邮　　编：** 100084
社 总 机： 010-83470000　　**邮　　购：** 010-62786544
投稿与读者服务： 010-62776969，c-service@tup.tsinghua.edu.cn
质量反馈： 010-62772015，zhiliang@tup.tsinghua.edu.cn
印 装 者： 北京鑫海金澳胶印有限公司
经　　销： 全国新华书店
开　　本： 185mm×260mm　　**印　　张：** 19.25　　**字　　数：** 347 千字
版　　次： 2025 年 8 月第 1 版　　**印　　次：** 2025 年 8 月第 1 次印刷
定　　价： 50.00 元

产品编号：110312-01

编 委 会

前　言

中学数学竞赛自20世纪中叶在我国广泛开展以来，已成为激发学生创新思维、培养逻辑推理和实践能力的重要平台。伴随着教育改革与国际交流的不断深入，越来越多的中国学生走上了国际数学竞赛的舞台，如美国数学竞赛（AMC）、澳大利亚数学竞赛、欧洲青少年奥林匹克竞赛等。这些国际数学竞赛不仅考查学生扎实的数学基础，更关注学生的创新思维、分析能力和英语学术表达能力。与此同时，当前国内市场上针对国际数学竞赛的系统性教程仍相对匮乏，难以满足广大学生和教师的实际需求。

为顺应当前国际数学教育发展的新趋势，翰林教育教研组组织一线教师和竞赛研究者，结合自身多年的竞赛教学实践，精心编写了《国际数学竞赛辅导（初级）》。本书以初等数论、初等代数、初等几何和组合数学为核心模块，全面涵盖国际主流数学竞赛中常见的重要知识点和题型。内容结构科学、体系完备，难度设置合理，既适合零基础学生循序渐进地入门学习，也满足有一定竞赛经验的学生深入提高的需求。

本书最大特色在于采用中英文对照形式编写。所有理论知识、例题解析及习题解答均附有英文原文及中文注释，既方便国内学生理解知识点、掌握解题技巧，也帮助他们逐步熟悉英文竞赛环境中的专业术语和学术表达。同时，这一设计也便于国际学校、双语学校以及有志于参加国际数学竞赛的学生更好地适应双语学习的需求，为今后在全球范围内参加各类数学赛事打下坚实的语言与学科基础。

本书注重“以例带学”，配备了大量经过遴选的经典例题与最新竞赛模拟题，题型丰富、难度递进，兼顾基础训练与综合应用，帮助学生在掌握知识的同时提升实际解题能力。针对不同阶段学生的特点，每章内容既有单一知识点的专项训练，也有多知识点融合的综合题目，力求培养学生灵活运用知识、独立分析和解决问题的能力。

在教学方法上，本书充分借鉴国际先进数学教育理念，强调问题驱动、归纳总结和多解法探索，鼓励学生主动思考、勇于创新。配套的练习题和拓展题目不仅满足日常训练，更能帮助学生在竞赛中获得实战经验。每一模块均配有详细的知识梳理与方法讲解，教师和教练可结合自身教学需求灵活取用，既能作为课堂讲授的教程，也可用于课后巩固与竞赛集训。

本书的编写团队汇聚了多位具有丰富竞赛辅导经验的老师，团队成员长期活跃于国内外一线教学和竞赛指导，熟悉当前各类主流竞赛的命题风格及趋势。我们深知，优质辅导教程是学生科学备赛、高效提升的关键保障。本书正是在对市场需求深入调研和多年教学积淀的基础上应运而生的，旨在为广大学生和教师提供系统、实用、前瞻性的国际数学竞赛学习资源。

我们衷心希望，本书能够成为每一位竞赛学生学习成长路上的良师益友，也能为广大教师、教练员的竞赛教学提供有力支持。愿所有热爱数学的同学在竞赛中不断挑战自我、收获成长，迈向更加广阔的学术殿堂。

编者

2025 年 3 月

目 录

Chapter 1　Factors and Prime Factorizations

第 1 章　因数与质因数分解

1.1　Basic Knowledge 基础内容

1.1.1　Factors 因数

Factors are the numbers you multiply together to get the product.

$1 \times 12 = 2 \times 6 = 3 \times 4 = 12 \Rightarrow$ 1, 2, 3, 4, 6 and 12 are factors of 12.

两个正整数相乘时，这两个数都叫作积的因数，或称为约数。

例如，$1 \times 12 = 12$，$2 \times 6 = 12$，$3 \times 4 = 12$。所以 1、2、3、4、6 和 12 都是 12 的因数。

1.1.2　Divisors 除数

A divisor is a number which divides the given number without leaving any remainder.

$$\frac{6}{1}=6,\quad \frac{6}{2}=3,\quad \frac{6}{3}=2,\quad \frac{6}{6}=1$$

The denominators 1, 2, 3, and 6 are called the divisors of the numerator 6.

Note: A factor of the number is the same as a divisor of the number.

除数是一个能将给定数整除且没有余数的数。

分母 1、2、3、6 是分子 6 的除数。

注意：一个数的因数与该数的除数是相同的。

1.1.3　Power 幂

A power is an integer that is equal to the product of a repeated multiplication of a number. The number of multplications is at least 2.

$2^5 = 32, 6^4 = 1296, 3^4 = 81, 5^6 = 15625$, are powers.

幂是一个数自乘若干次所得的积，重复连乘的运算叫作乘方，乘方的结果叫作幂。乘方运算至少为 2 次。

$2^5=32$，$6^4=1296$，$3^4=81$，$5^6=15625$，这些结果是幂。

1.1.4 Composite Number 合数

When a number has more than two factors (not including two), the number is called a composite number.

当一个数有两个以上（不含两个）的因数时，这个数称为合数。

1.1.5 Prime Number 质数

A prime number is a positive integer greater than 1 and only divisible by 1 and itself. Another way of saying this is that a prime number is a positive integer with exactly two factors (1 and itself).

一个大于 1，且除了 1 和它自身外，不能被其他自然数整除的自然数叫作质数，又称素数。即质数只有两个正因数（1 和它本身）。

Twenty-five prime numbers (up to 100) 100 以内的 25 个质数

2		3	5	7	
	11	13		17	19
		23			29
	31			37	
	41	43		47	
		53			59
	61			67	
	71	73			79
		83			89
				97	

1.1.6 The Smallest Prime Number 最小的质数

2, which is an even number and the only even prime number.

(**Note:** 1 is neither a prime number nor a composite number.)

2 是偶数，也是唯一的偶质数。

（**注意**：1 既不是质数也不是合数。）

1.2 Method To Determine A Prime Number 确定质数的方法

1.2.1 The Square Root Rule 平方根规则

If a is not divisible by all the prime numbers less than or equal to $\sqrt{a}$, then a is a prime number.

如果 a 不被所有小于或等于 $\sqrt{a}$ 的质数整除，则 a 是一个质数。

For example: 59 is a prime number, because 59 isn't divisible by any prime number smaller than $\sqrt{59}$ (which are: 2, 3, 5, 7).

例如：59 是一个质数，因为 59 无法被任何小于 $\sqrt{59}$ 的质数（即 2、3、5、7）整除。

Property 1 A prime number p can only be written as $p \times 1$.

性质 1 一个质数 p 只能表示为 $p \times 1$。

In other words, a prime number p can only be divided by itself and 1. If p is a prime number and $p = mn$, where m and n are positive integers, then one of the two numbers, m and n, must be 1 and the other must be p.

换句话说，质数 p 只能被 1 和它自身整除。如果 p 是一个质数且 $p=mn$，并且 m, n 均是正整数，那么 m 和 n 中必有一个是 1，另一个是 p。

Property 2 If the sum of two prime numbers is an odd number, then one of the two prime numbers must be 2.

性质 2 如果两个质数的和是一个奇数，那么这两个质数中必有一个是 2。

Property 3 If the product of two prime numbers is an even number, then one of the two prime numbers must be 2.

性质 3 如果两个质数的积是一个偶数，那么这两个质数中必有一个是 2。

1.2.2 Relatively Prime 互质

If two positive integers have no common factor except 1, they are said to be relatively prime. For example, 4 and 9 are relatively prime.

如果两个正整数除了 1 之外没有其他公因数，则称这两个正整数互质。例如，4

和 9 互质。

Example 1

The sum of two prime numbers is 91. What is the product of these two prime numbers?

例题 1

两个质数的和为 91。这两个质数的积是多少？

(A) 86　　(B) 98　　(C) 125　　(D) 156　　(E) 178

Solution

According to Property 2, since the sum of the two prime numbers is an odd number, one of them must be even. The only even prime number is 2. The other prime number is $91 - 2 = 89$, and the product of these two numbers is $89 \times 2 =$ (E)178.

解　析

根据性质 2，由于这两个质数的和是一个奇数，那么其中一个必须是偶数，唯一的偶质数是 2。另一个质数是 $91-2=89$，因此这两个数的乘积是 $89 \times 2=178$，故选 (E)。

Example 2

Three members of the Hanlin Middle School girls' football team had the following conversation:

Ashley: I just realized that our uniform numbers are all 2-digit primes.

Bethany: And the sum of your two uniform numbers is the date of my birthday later this month.

Caitlin: That's funny. The sum of your two uniform numbers is the date of my birthday earlier this month.

Ashley: And the sum of your two uniform numbers is today's date.

What number does Ashley wear?

例题 2

翰林中学女子足球队的三名成员进行了如下对话：

艾什莉：我刚刚意识到我们的球衣号码都是两位数的质数。

贝瑟尼：而且你们两个球衣号码的和是我这个月即将到来的生日的日期。

凯特琳：更有趣的是，你们两个球衣号码的和是我这个月刚过完的生日的日期。

艾什莉：而且你们两个球衣号码的和是今天的日期。

请问艾什莉穿的是几号球衣？

(A) 11 (B) 13 (C) 17 (D) 19 (E) 23

Solution

Since the maximum number of days in a month is 31, the three uniform numbers should be 11, 13 and 17, the three smallest 2-digit prime numbers. Otherwise, one of the sums of two numbers would be greater than 31. Thus, the sums are 24, 30 and 28, all representing three days.

Since Bethany says that the sum of other two people's numbers is a later date, the numbers of Caitlin and Ashley must add up to 30. Since Caitlin says that the sum of other two people's numbers is earlier, it must be 24. This leaves 28 as today's date. Therefore, Bethany's number and Caitlin's number add up to 28, and Ashley's number is (B)13.

解 析

一个月最多有 31 天，因此三个球衣号码必须是最小的三个两位数质数 11、13 和 17，否则就会有两个号码的和大于 31。所以号码的和为 24、28 和 30，也代表这个月中的 3 天。

由于贝瑟尼说其他两人的号码和所代表的日期还没到，所以凯特琳和艾什莉的号码相加为 30。同样，凯特琳说其他两个人的号码和所代表的日期已经过了，所以只能是 24。因此艾什莉说的“今天的日期”是 28，这意味着贝瑟尼和凯特琳的号码加起来是 28，所以艾什莉的号码是 13，答案是 (B)。

1.3 Fundamental Theorem of Arithmetic 算术基本定理

Fundamental Theorem of Arithmetic states that all the natural numbers greater than 1 can be expressed in the form of a product of its prime factors. And this expression is unique.

算术基本定理指出，所有大于 1 的自然数都可以表示成其质因数相乘的形式。且这种表示是唯一的。

$36 = 6 \times 6 = (2 \times 3) \times (2 \times 3) = 2 \times 2 \times 3 \times 3 = 2^2 \times 3^2$,

$36 = 4 \times 9 = (2 \times 2) \times (3 \times 3) = 2 \times 2 \times 3 \times 3 = 2^2 \times 3^2$,

$36 = 3 \times 12 = 3 \times (2 \times 6) = 3 \times (2 \times 2 \times 3) = 2 \times 2 \times 3 \times 3 = 2^2 \times 3^2$,

$36 = 2 \times 18 = 2 \times (2 \times 9) = 2 \times (2 \times 3 \times 3) = 2 \times 2 \times 3 \times 3 = 2^2 \times 3^2$,

$36 = 1 \times 36 = 1 \times (2 \times 3)^2 = 2^2 \times 3^2$。

1.3.1 Prime Factorizations 质因数分解

Prime factorization of a number is to express the number as a product of factors that are all primes. (A prime can also be factorized, and the result is itself.)

$12 = 2 \times 6 = 2 \times 2 \times 3 = 2^2 \times 3$.　　$1001 = 7 \times 11 \times 13$.

把一个数分解成若干个质因数乘积的形式，即求质因数的过程叫作分解质因数。(质数也可以进行质因数分解，结果就是它本身。)

$12 = 2 \times 6 = 2 \times 2 \times 3 = 2^2 \times 3$。　　$1001 = 7 \times 11 \times 13$。

1.3.2 Perfect Square 完全平方数

A perfect square is an integer that is equal to some integer to the second power.

0，1，4，9，16，25 are perfect squares.

完全平方数是某个整数自乘两次后所得的整数，即该整数的二次方。

0，1，4，9，16 和 25 是完全平方数。

1.3.3 Perfect Cube 完全立方数

A perfect cube is an integer that is equal to some integer to the third power.

0，1，8，27，64，125 are perfect cubes.

完全立方数是某个整数自乘三次后所得的整数，即该整数的三次方。

0，1，8，27，64 和 125 是完全立方数。

Example 3

Which of the following numbers is not a perfect square?

例题 3

以下哪个选项不是完全平方数？

(A) 1^{4016}　　(B) 2^{4017}　　(C) 3^{4018}　　(D) 4^{4019}　　(E) 5^{4040}

Solution

To tell if a power is a square or not, we first check the exponent. If the exponent is even, then it is a square, leaving (B) and (D) to be the candidates for the answer. Notice that 4 itself is a square. According to the exponentiation law, we know 4^{4019} is a square. The answer is (B)2^{4017}.

解　析

判断一个数是否为完全平方数，先看它的指数。如果指数是偶数，则它是完全平方数，因此剩下选项 (B) 和 (D)。请注意 4 本身就是一个完全平方数。根据乘方运算法则，4^{4019} 是一个完全平方数，因此只有 2^{4017} 不是完全平方数，答案是 (B)。

Example 4

The positive integers x and y are the two smallest positive integers for which the product of 120 and x is a square and the product of 120 and y is a cube. What is the sum of x and y?

例题 4

正整数 x 和 y 是满足 120 与 x 的乘积是完全平方数，且 120 与 y 的乘积是完全立方数的两个最小正整数。求 x 和 y 的和。

(A) 85　　(B) 95　　(C) 105　　(D) 155　　(E) 225

Solution

If a number is a perfect square, all of the exponents in its prime factorization must be even. The prime factorization of 120 is $2^3 \times 3 \times 5$. To make $120x$ a perfect square and minimize x, we can multiply a 2, a 3 and a 5 with 120, so the value of x is 30. Similarly, to have a perfect cube, the exponents in the prime factorization must be multiples of 3. Thus the value of y is $3^2 \times 5^2 = 225$. Therefore, our answer is (E)225.

解　析

如果一个数是完全平方数，必须保证质因数分解中的所有指数都是偶数。120 的质因数分解为 $120 = 2^3 \times 3 \times 5$。因此，至少需要给 120 乘以一个 2、一个 3 和一个 5，才能使 $120x$ 成为完全平方数，因此 x 的值为 30。同样，一个数是完全立方数等价于质因数分解中的指数都是 3 的倍数。因此 y 的值是 $3^2 \times 5^2 = 225$。所以答案是 (E)。

1.3.4 Number of divisors 因数个数

For an integer n greater than 1, let the prime factorization of n be。

$$n = p_1^{a_1} p_2^{a_2} p_3^{a_3} \cdots p_k^{a_k}$$

where $a_1, a_2, a_3, \cdots, a_k$ are nonnegative integers, $p_1, p_2, p_3, \cdots, p_k$ are prime numbers.

The number of divisors: $d(n) = (a_1 + 1)(a_2 + 1)(a_3 + 1)\cdots(a_k + 1)$.

对于一个大于 1 的整数 n，设 n 的质因数分解为

$$n = p_1^{a_1} p_2^{a_2} p_3^{a_3} \cdots p_k^{a_k}$$

其中 a_1，a_2，a_3，⋯，a_k 是非负整数，p_1，p_2，p_3，⋯，p_k 是质数。

因数个数： $d(n) = (a_1 + 1)(a_2 + 1)(a_3 + 1)\cdots(a_k + 1)$。

Example 5

How many positive factors does 44352 have?

例题 5

计算 44352 有多少个因数?

(A) 42　　(B) 54　　(C) 66　　(D) 78　　(E) 84

Solution

We can first find the prime factorization of 44352, which is $2^6\times3^2\times7\times11$. Now, we add one to our exponents and multiply them together. Therefore, the answer is $(6+1)(2+1)\cdot(1+1)(1+1)=$ (E)84.

解 析

先将 44352 做质因数分解，得到 $2^6\times3^2\times7\times11$。将指数加 1 并相乘，得到答案 $(6+1)(2+1)(1+1)(1+1)=84$，故选 (E)。

Example 6

For any positive integer n, (n) is the sum of the positive factors of n. For example, $(6)=1+2+3+6=12$. Find $((13))$.

例题 6

对任意正整数 n, (n) 表示 n 所有正因数的和，例如，$(6)=1+2+3+6=12$。求 $((13))$。

(A) 13 (B) 20 (C) 24 (D) 28 (E) 30

Solution

According to the definition of (n), we have

$(13)=1+13=14$,

$((13))=(14)=1+2+7+14=24$.

Thus, the answer is (C).

解 析

根据 (n) 的定义，得

$(13)=1+13=14$,

$((13))=(14)=1+2+7+14=24$。

因此，答案是 (C)。

1.4 GCF and LCM 最大公因数和最小公倍数

1.4.1 The Greatest Common Factor 最大公因数

The **greatest common factor** (GCF) of a group of natural numbers is the greatest natural number that is a factor of every number in the group.

最大公因数（简称 GCF）是一个自然数，它是组内所有数共同的因数中最大的那个。

Properties of the Greatest Common Factor:

If $(a, b) = d$, and n is a positive integer, then $(na, nb) = nd$.

最大公因数的性质：

如果 $(a, b)=d$，且 n 是正整数，则 $(na, nb)=nd$。

$(a, 1) = 1$; $(a, a) = a$; $(a, 0) = a$; $(a, b) = (b, a)$; $(a, b) = (b, a - b)$。

1.4.2 The Least Common Multiple 最小公倍数

The least common multiple (LCM) of a group of natural numbers is the smallest natural number that is a multiple of every number in the group.

最小公倍数（简称 LCM）是一个自然数，它是组内所有数共同的倍数中最小的那个。

Notation:

Given two positive integers a and b, the greatest common factor of a and b is written as (a, b) and the least common multiple of a and b is written as $[a, b]$.

符号：

给定两个正整数 a 和 b，a 和 b 的最大公因数记作 (a, b)，a 和 b 的最小公倍数记作 $[a, b]$。

1.4.3 Finding The Greatest Common Factor 如何计算最大公因数

Step 1: Find the prime factorization of each number.

Step 2: Find all primes common to all numbers.

Step 3: Find all terms with the smallest exponent common to all factorizations.

Step 4: Multiply all terms in Step 3 and the product is the greatest common factor.

步骤 1：对每个数进行质因数分解。

步骤 2：找出所有数共有的质因数。

步骤 3：针对每一个质因数，找出同一个质因数中指数最小的那一项。

步骤 4：将步骤 3 中的所有质因数相乘，得到的积就是最大公因数。

1.4.4 Finding the Least Common Multiple 如何计算最小公倍数

Step 1: Find the prime factorization of each number.

Step 2: Find all primes belonging to any factorization.

Step 3: Find all terms with the largest exponent in any factorization.

Step 4: Multiply all terms in Step 3 and the product is the least common multiple.

步骤 1：对每个数进行质因数分解。

步骤 2：找出所有质因数（包含在任何一个分解中的质数）。

步骤 3：针对每一个质因数，选出同一个质因数中指数最大的那一项。

步骤 4：将步骤 3 中的所有质因数相乘，得到的积就是最小公倍数。

Relationship between LCM and GCF 最小公倍数与最大公因数的关系

$$(\text{LCM of } a \text{ and } b) \times (\text{GCF of } a \text{ and } b) = a\times b,$$

$$(a \text{ 和 } b \text{ 的最小公倍数}) \times (a \text{ 和 } b \text{ 的最大公因数}) = a\times b,$$

$$[\boldsymbol{a}, \boldsymbol{b}] \times (\boldsymbol{a}, \boldsymbol{b}) = \boldsymbol{ab}.$$

Example 7

What is the difference between the least common multiple of 60 and 102 and the greatest common factor of 12 and 30?

例题 7

60 和 102 的最小公倍数与 12 和 30 的最大公因数的差是多少？

(A) 60 (B) 6120 (C) 1020 (D) 1014 (E) 720

Solution

We do the prime factorizations first.

$60=2^2\times 3\times 5$, $102=2\times 3\times 17$.

Thus, $[60, 102] =2^2\times 3\times 5\times 17=1020$.

Similarly, we have $12=2^2\times 3$, $30=2\times 3\times 5$.

So, $(12, 30)=2\times 3=6$.

1020−6=(D)1014.

解 析

先进行质因数分解。

$60=2^2\times3\times5$，$102=2\times3\times17$。

所以 60 和 102 的最小公倍数为 $2^2\times3\times5\times17=1020$。

同样地，质因数分解可得 $12=2^2\times3$，$30=2\times3\times5$。

所以 12 和 30 的最大公因数为 $2\times3=6$。

1020−6=1014，故答案是 (D)。

Example 8

The smallest number greater than 2 that leaves a remainder of 1 when divided by 3, 4, 5, or 6 lies between which two numbers?

例题 8

大于 2 且在被 3、4、5 或 6 除时余数为 1 的最小数字在哪两个数之间?

(A) 30 and 39　　30 和 39

(B) 40 and 49　　40 和 49

(C) 50 and 59　　50 和 59

(D) 60 and 69　　60 和 69

(E) 70 and 79　　70 和 79

Solution

To find the answer to this problem, we need to find the least common multiple of 3, 4, 5, 6 and add 1 to the result. The least common multiple of the four numbers is 60, and by adding 1, we find that the number is 61. Now we need to find the range that contains 61. The only such a range is in answer (D).

解 析

要找到这个问题的答案，先找 3、4、5 和 6 的最小公倍数，然后在结果上加上 1。这四个数的最小公倍数是 60，加上 1，我们发现这个数是 61。现在我们需要找出包含

61 的区间。唯一这样的区间的答案是 (D)。

Example 9

The least common multiple of a and b is 15, and the least common multiple of b and c is 35. What is the least possible value of the least common multiple of a and c?

例题 9

a 和 b 的最小公倍数是 15，b 和 c 的最小公倍数是 35。a 和 c 的最小公倍数的最小可能值是多少？

(A) 21　(B) 24　(C) 25　(D) 65　(E) 105

Solution

Since the LCM of a and b is 15, we know a and b are factors of 15; similarly, b and c are factors of 35. To minimize the LCM of a and c, we can maximize b. Since b is a common factor of 15 and 35, its maximum value is 5. In this case, $a = 3$, $c = 7$, and the LCM of a and c is (A)21.

解　析

由于 a 和 b 的最小公倍数为 15，可知 a 和 b 是 15 的因子；同理，b 和 c 是 35 的因子。为了使 a 和 c 的最小公倍数最小，我们让 b 的值尽量大。由于 b 是 15 和 35 的公因子，因此它的最大值为 5。这种情况下，$a = 3$，$c = 7$，a 和 c 的最小公倍数是 21，故选 (A)。

Example 10

What is the ratio of the least common multiple of 180 and 1320 to the greatest common factor of 180 and 1320?

例题 10

180 和 1320 的最小公倍数与 180 和 1320 的最大公因数的比是多少？

(A) 11　(B) 22　(C) 66　(D) 99　(E) 180

Solution

To find either the LCM or the GCF of two numbers, we do prime factorization first:

The prime factorization of 180 is $2^2 \times 3^2 \times 5$.

The prime factorization of 1320 is $2^3 \times 3 \times 5 \times 11$.

Hence, $[180,1320] = 2^3 \times 3^2 \times 5 \times 11$ and $(180,1320) = 2^2 \times 3 \times 5$. Therefore, $\dfrac{2^3 \times 3^2 \times 5 \times 11}{2^2 \times 3 \times 5} = 2 \times 3 \times 11 = 66$, so the answer is (C).

解　析

要找到两个数的最小公倍数和最大公因数，先进行质因数分解。

180 的质因数分解为 $180 = 2^2 \times 3^2 \times 5$。

1320 的质因数分解为 $1320 = 2^3 \times 3 \times 5 \times 11$。

因此，这两个数的最小公倍数为 $2^3 \times 3^2 \times 5 \times 11$，且这两个数的最大公因数是 $2^2 \times 3 \times 5$。

因此，$\dfrac{2^3 \times 3^2 \times 5 \times 11}{2^2 \times 3 \times 5} = 2 \times 3 \times 11 = 66$，可得答案为 (C)。

PRACTICE 练习

Problem 1

The digits 2, 0, 2 and 5 are placed in the expression below, one digit per box. What is the maximum possible value of the expression?

问题 1

数字 2，0，2 和 5 被放置在下列表达式中，每个框中一个数字。这个表达式的最大可能值是多少？

$$\square^{\square} \times \square^{\square}$$

(A) 0　　(B) 4　　(C) 16　　(D) 25　　(E) 32

Problem 2

Which of the following numbers has the smallest prime factor?

问题 2

以下哪个数有最小的质因数？

(A) 75 (B) 77 (C) 78 (D) 79 (E) 81

Problem 3

What is the smallest positive integer that is neither prime nor square and that has no prime factor less than 40?

问题 3

既不是质数也不是平方数，并且没有小于 40 的质因数的最小正整数是什么数？

(A) 1517 (B) 1763 (C) 1767 (D) 1913 (E) 1947

Problem 4

In how many ways can 10003 be written as the sum of two primes?

问题 4

10003 可以用几种方式写成两个质数的和？

(A) 0 (B) 1 (C) 2 (D) 3 (E) 4

Problem 5

Jerry chose a four-digit number to be the personal identification number for his bank account. The first digit (leftmost digit) is prime, the greatest common factor of the middle two digits is 3, and the last digit is a divisor of 20. Which of the following numbers could be his personal identification number?

问题 5

杰瑞用一个四位数作为他的银行账户的个人识别号码。第一个数字（最左侧数字）是质数，中间两个数字的最大公因数是 3，最后一个数字是 20 的一个因数。以下哪个数字可能是他的个人识别号码？

(A) 2363 (B) 3465 (C) 3964 (D) 5647 (E) 6396

Problem 6

Find a positive integer such that the sum of the positive integer and 57 is a square number, and the sum of the positive integer and 74 is also a square number.

问题 6

找到一个正整数使得该数与 57 的和是一个平方数，且该数与 74 的和也是一个平方数。

(A) 8　(B) 9　(C) 7　(D) 17　(E) 24

Problem 7

For three positive prime numbers a, b, and c, we know $ab = 21$ and $bc = 77$. What is the value of abc?

问题 7

三个正质数 a, b, c, 其中 $ab = 21$ 且 $bc = 77$。那么 abc 的积是多少？

(A) 121　(B) 231　(C) 263　(D) 363　(E) 1617

Problem 8

How many 2-digit numbers have digits whose sum is a perfect cube?

问题 8

有多少个两位数，其数位上的数字之和是一个完全立方数？

(A) 7　(B) 8　(C) 9　(D) 10　(E) 11

Problem 9

What is the sum of the distinct prime integer divisors of 2025?

问题 9

2025 的不同质因数的和是多少？

(A) 5　(B) 7　(C) 8　(D) 14　(E) 63

Problem 10

Barry wrote 6 different numbers, one on each side of 3 cards, and laid the cards on a table, as shown. The sums of the two numbers on each of the three cards are equal. The three numbers on the hidden sides are prime numbers. What is the average of the hidden prime numbers?

问题 10

巴里在 3 张卡片的每一面上分别写了 6 个不同的数字，并将这些卡片摆放在桌子上，如图所示。每张卡片上正反两个数字的和与其他两张卡片相等。隐藏面上的三个数字是质数。隐藏的质数的平均值是多少？

(A) 6　(B) 12　(C) 15　(D) 16　(E) 17

40 46 49

Problem 11

Let w, x, y, and z be positive integers. If $2^w \cdot 3^x \cdot 5^y \cdot 7^z = 360$, then what does $2w+3x+5y+7z$ equal?

问题 11

设w，x，y，z都是正整数。如果$2^w \cdot 3^x \cdot 5^y \cdot 7^z=360$，那么$2w+3x+5y+7z$等于多少？

(A) 7 (B) 17 (C) 27 (D) 35 (E) 56

Problem 12

Let N be the smallest five-digit number whose digits have a product of 120. What is the sum of the digits of N?

问题 12

设N是每个数位数字乘积为 120 的最小的五位数。那么N的每个数位数字之和是多少？

(A) 15 (B) 16 (C) 17 (D) 18 (E) 20

Problem 13

On June 1, a group of students are standing in rows, with 15 students in each row. On June 2, the same group is standing with all of the students in one long row. On June 3, the same group is standing with just one student in each row. On June 4, the same group is standing with 6 students in each row. This process continues through June 12 with a different number of students per row each day. However, on June 13, they cannot find a new way of organizing the students. What is the smallest possible number of students in the group?

问题 13

在 6 月 1 日，一组学生排成若干排，每排 15 名学生。6 月 2 日，同一组学生排成一长排。6 月 3 日，同一组学生每排只有 1 名学生。6 月 4 日，同一组学生每排 6 名学生。这个过程一直持续到 6 月 12 日每天每排的学生数不同。然而，到了 6 月 13 日，他们找不到新的排列方式。这个组中最少可能有多少名学生？

(A) 21 (B) 36 (C) 60 (D) 90 (E) 445

Problem 14

I'm thinking of two positive integers. Their product is 24 and their difference is 5. What is the smaller number?

问题 14

我在想两个正整数，它们的乘积是 24，差是 5。较小的那个数字是多少？

(A) 3　(B) 4　(C) 6　(D) 8　(E)12

Problem 15

How many integers between 1000 and 3000 have all three of the numbers 15, 20, and 25 as factors?

问题 15

在 1000 和 3000 之间有多少个整数同时是 15，20 和 25 的倍数？

(A) 1　(B) 3　(C) 5　(D) 7　(E) 9

Problem 16

Find the least common multiple of 21, 24, and 42.

问题 16

21，24，42 的最小公倍数是多少？

(A) 106　(B) 126　(C) 168　(D) 204　(E) 408

Problem 17

The figure shows parts of two circular gears whose teeth interlock when the gears turn. Gear A has 84 teeth and gear B has 56 teeth. What is the least positive number of rotations gear B must make before the arrows will again be aligned in the same way?

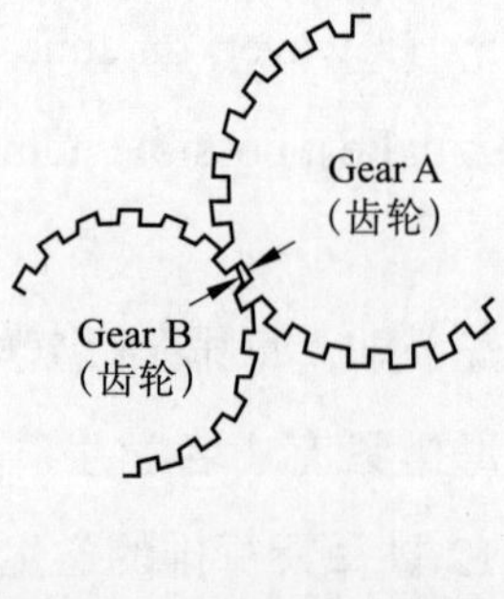

问题 17

图中显示了两个圆形齿轮的部分，它们的齿轮在转动时相互啮合。齿轮 A 有 84 个齿，齿轮 B 有 56 个齿，齿轮 B 最少必须转动几圈才能使箭头再次对齐？

(A) 1　(B) 2　(C) 3　(D) 7　(E) 5

Chapter 2 Odd, Even and Divisibility

第 2 章 奇数、偶数和整除

2.1 Odd and Even 奇数和偶数

2.1.1 Even Integer 偶数

An integer with the last digit of 0, 2, 4, 6, or 8. General form: $2n$ or $2n+2$, where n is any integer.

Examples: Even integers: 10, 12, 14, 16, and 18.

偶数是一个整数，其末位数字为 0，2，4，6 或 8，其通式为 $2n$ 或 $2n+2$，其中 n 为任意整数。

例如：10，12，14，16 和 18 为偶数。

2.1.2 Odd Integers 奇数

An integer with the last digit of 1, 3, 5, 7, or 9. General form: $2n+1$ or $2n-1$.

Examples: Odd integers: 11, 13, 15, 17, and 19.

奇数是一个整数，其末位数字为 1，3，5，7 或 9，其通式为 $2n+1$ 或 $2n-1$。

例如：11，13，15，17 和 19 为奇数。

2.1.3 Parity 奇偶性

An even number has even parity and an odd number has odd parity.

偶数具有偶性，奇数具有奇性。

Properties 性质

even ± even = even	偶数 ± 偶数 = 偶数	$2+2=4, 4-2=2$
even ± odd = odd	偶数 ± 奇数 = 奇数	$2+1=25, 3-2=1$

odd ± odd = even	奇数 ± 奇数 = 偶数	$1+1=26, 3-1=2$
odd × odd = odd	奇数 × 奇数 = 奇数	$15 \times 15=225$
odd ÷ odd = odd	奇数 ÷ 奇数 = 奇数	$1001 \div 11=91$
(without remainder)	(可整除时)	
odd × even = even	奇数 × 偶数 = 偶数	$11 \times 12=132$
odd ≠ even	奇数 ≠ 偶数	$1 \neq 2$

Example 1

Suppose m and n are positive odd integers. Which of the following must also be an odd integer?

例题 1

假设 m 和 n 是正奇数。以下哪个表达式也是奇数?

(A) $m+5n$　(B) $5m-n$　(C) $5m^2+3n^2$　(D) $(nm+7)^2$　(E) $11mn$

Solution

Assume that m and n are both 1. Plugging into each of the choices, we get (E) $11mn$ is an odd integer.

解　析

假设 m 和 n 都是 1,将其代入每个选项中,可得到 $11mn$ 是奇数。答案是 (E)。

Example 2

If a and b are integers and a^2+b^2 is even, which of the following is impossible?

例题 2

如果 a 和 b 是整数且 a^2+b^2 是偶数,那么以下哪个情况是不可能的?

(A) a and b are even　a 和 b 都是偶数

(B) a and b are odd　a 和 b 都是奇数

(C) $a+b$ is even　$a+b$ 是偶数

(D) $a+b$ is odd　　$a+b$ 是奇数

(E) none of these are impossible　　以上都可能

Solution

Since a^2+b^2 is even, either both a^2 and b^2 are even, or they are both odd. Therefore, a and b are either both even or both odd, since the square of an even number is even and the square of an odd number is odd. As a result, $a+b$ must be even. The answer, then, is (D).

解 析

由于 a^2+b^2 是偶数，因此 a^2 和 b^2 要么都是偶数，要么都是奇数。因此，a 和 b 要么都是偶数，要么都是奇数，因为偶数的平方是偶数，奇数的平方是奇数，所以 $a+b$ 一定是偶数。所以答案是 (D)。

2.2 Problem Solving Skills 解题技巧

- The sum of any even integer and 1 is odd: $4+1=5$.
- The sum of two consecutive integers is odd: $n+(n+1)=2n+1$; $12+13=25$.
- The product of two consecutive integers is even: $n(n+1)$ is even; $12 \times 13=156$.
- Any two consecutive integers have opposite parity: 12 is even and 13 odd.
- $a+b$ and $a-b$ has the same parity: $15-5=10$; $15+5=20$.
- If the product of n positive integers is even, at least one of these n positive integers is even.
- If the product of n positive integers is odd, all of these n positive integers are odd.
- If the number of odd integers is even, then the sum of them is even.
- If the number of odd integers is odd, then the sum of them is odd.
- 任意偶数与 1 的和是奇数：$4+1=5$。
- 两个连续整数的和是奇数：$n+(n+1)=2n+1$；$12+13=25$。
- 两个连续整数的积是偶数：$n(n+1)$ 是偶数；$12\times 13=156$。
- 任意两个连续整数具有相反的奇偶性：12 是偶数，13 是奇数。
- $a+b$ 和 $a-b$ 具有相同的奇偶性：$15-5=10$；$15+5=20$。

- 如果 n 个正整数的积是偶数，至少有一个正整数是偶数。
- 如果 n 个正整数的积是奇数，则所有这些正整数都是奇数。
- 如果有偶数个数的奇数，那么它们的和是偶数。
- 如果有奇数个数的奇数，那么它们的和是奇数。

Example 3

Which of the following integers cannot be written as the sum of four consecutive even integers?

例题 3

以下哪个整数不能表示为四个连续偶数的和?

(A) 20　　(B) 40　　(C) 44　　(D) 52　　(E) 60

Solution

Let our 4 numbers be n, $n + 2$, $n + 4$, $n + 6$, where n is even. Then our sum is $4n + 12$. The only answer choice that cannot be written as $4n + 12$, where n is even, is (B)40.

解　析

设这四个数为 n、$n+2$、$n+4$、$n+6$，其中 n 是偶数。那么它们的和为 $4n+12$。唯一一个不能表示为 $4n+12$（n 是偶数）的选项是 (B)。

Example 4

When a positive integer N is fed into a machine, the output is a number calculated according to the rule shown below.

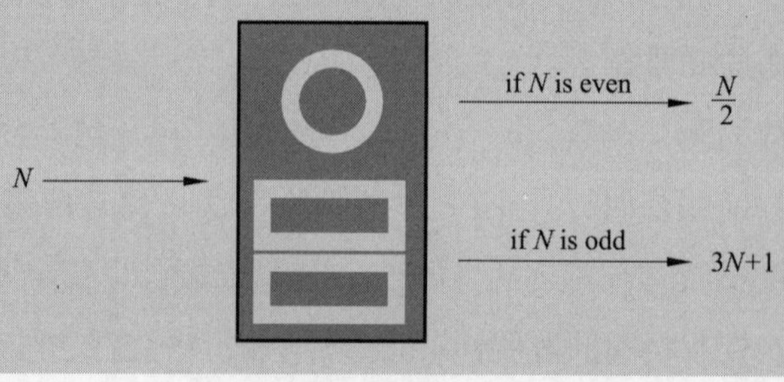

For example, starting with an input of N=7 the machine will output $3\times7+1=22$. Then if the output is repeatedly inserted into the machine five more times, the final output is 26.

$$7\to22\to11\to34\to17\to52\to26$$

When the same 6-step process is applied to a different starting value of N the final output is 2. What is the sum of all such integers N?

$$N\to\_\_\_\_\to\_\_\_\_\to\_\_\_\_\to\_\_\_\_\to\_\_\_\_\to2$$

例题 4

当一个正整数输入到一个机器中时，输出是根据下面的规则计算的。

例如，假设输入 N=7，机器将输出 $3\times7+1=22$。然后，如果将这个输出结果重复输入机器五次，最终的输出是 26。

$$7\to22\to11\to34\to17\to52\to26$$

当相同的 6 步过程应用于不同的起始值 N 时，最终的输出是 2。所这样的整数 N 的总和是多少？

$$N\to\_\_\_\_\to\_\_\_\_\to\_\_\_\_\to\_\_\_\_\to\_\_\_\_\to2$$

(A) 40 (B) 128 (C) 130 (D) 170 (E) 183

Solution

We start with final output of 2 and work backward, by considering all possible inputs that could give the particular output. This produces following set of possibilities each stage:

(2) → (4) → (1, 8) → (2, 16) → (4, 5, 32) → (1, 8, 10, 64) → (2, 3, 16, 21, 128)

So the sum of all integers N is $2+3+16+21+128=$(D)170.

解 析

从最终结果 2 开始反推，考虑所有可能输入的数字，将产生以下每次计算输入的可能性集合：

(2) → (4) → (1, 8) → (2, 16) → (4, 5, 32) → (1, 8, 10, 64) → (2, 3, 16, 21, 128)

因此，所有这样的整数 N 的总和是 $2+3+16+21+128=170$，故选 (D)。

2.3 Divisibility 整除

"Divisible" means "when you divide one number by another number, the result is a whole number".

"Divisible by" and "can be evenly divided by" mean the same thing.

"整除"指的是"当你用一个数去除另一个数时，结果是整数"。

"被……整除"与"可以被……整除"含义相同。

2.3.1 Divisibility Rule For 2, 4 and 8 可以被 2, 4, 8 整除的规则

- A number is divisible by 2 if the last digit of the number is divisible by 2 (2^1).
- A number is divisible by 4 if the last two digits of the number form a number that is divisible by 4 (2^2).
- A number is divisible by 8 if the last three digits of the number form a number that is divisible by 8 (2^3).
- 一个数如果末位数字可以被 2 整除，则这个数可以被 2 整除（即 2^1 情形）。
- 一个数如果末两位数字形成的数可以被 4 整除，则这个数可以被 4 整除（即 2^2 情形）。
- 一个数如果末三位数字形成的数可以被 8 整除，则这个数可以被 8 整除（即 2^3 情形）。

2.3.2 Divisibility Rule For 5, 25 and 125 可以被 5, 25, 125 整除的规则

- A number is divisible by 5 if the last digit of the number is divisible by 5 (5^1).
- A number is divisible by 25 if the last two digits of the number form a number that is divisible by 25 (5^2).
- A number is divisible by 125 if the last three digits of the number form a number that is divisible by 125 (5^3).
- 一个数如果末位数字可以被 5 整除，则这个数可以被 5 整除（即 5^1 情形）。
- 一个数如果末两位数字形成的数可以被 25 整除，则这个数可以被 25 整除（即 5^2 情形）。

- 一个数如果末三位数字形成的数可以被 125 整除，则这个数可以被 125 整除（即 5^3 情形）。

2.3.3 Divisibility Rule For 3 and 9 可以被 3 和 9 整除的规则

- A number is divisible by 3 if the sum of the digits of the number is divisible by 3.
- A number is divisible by 9 if the sum of the digits of the number is divisible by 9.
- 一个数如果其各个数位数字之和可以被 3 整除，则这个数可以被 3 整除。
- 一个数如果其各个数位数字之和可以被 9 整除，则这个数可以被 9 整除。

Example 5

The 7-digit numbers 47A51B2 and 36A2BC4 are each multiple of 3. Which of the following could be the value of C?

例题 5

7 位数 47A51B2 和 36A2BC4 都是 3 的倍数。以下哪一个可能是 C 的值？

(A) 1　　(B) 2　　(C) 3　　(D) 5　　(E) 8

Solution

Since both numbers are divisible by 3, the sum of their digits has to be divisible by 3. In the first number, we have $4+7+5+1+2 = 19$. In order to be a multiple of 3, $A + B$ has to be either 2 or 5 or 8 and so on. We add up the numerical digits in the second number; $3+6+2+4 = 15$. We then add two of the selected values, 5 to 15, to get 20. We then see that $C = 1$, 4 or 7, 10 and so on, otherwise the number will not be divisible by three. We then add 8 to 15, to get 23, which shows us that $C = 1$ or 4 or 7 and so on. In order to be a multiple of three, we select a few of the common numbers, which could be 1, 4, and 7. However, in the answer choices, there is no 7 or 4 or anything greater than 7, but there is a 1. (A) is your answer.

解　析

由于这两个数都能被 3 整除，它们的数字之和一定都能被 3 整除。先将第一个数各个数位上的已知的数字加起来得到 $4+7+5+1+2=19$。因为 19 除以 3 的余数为 1，所以 A+B 除以 3 的余数为 2。然后我们将第二个数字中已知的数字加起来得到

3+6+2+4=153。由于 15 是 3 的倍数，所以 A+B+C 也是 3 的倍数。考虑到 A+B 的余数，可知 C 除以 3 的余数为 1。由于 C 的取值在 0 ～ 9 之间，所以 C 可能是 1、4 和 7。故答案是 (A)。

Example 6

Find the least possible value of digit d so that 7340d3 is divisible by 9.

例题 6

在下列选项中找到最小的数字 d，使得 7340d3 可被 9 整除。

(A) 1　　(B) 2　　(C) 3　　(D) 4　　(E) 5

Solution

First, an integer is divisible by 9 if and only if the sum of its digits is divisible by 9. So, we add all the digits together and get $7+3+4+d+0+3=17+d$. To have $17+d$ a multiple of 9, (A) 1 is the only option.

解　析

被 9 整除的规则是所有数位之和是 9 的倍数，因此将该数各数位上的数字相加 $7+3+4+d+0+3=17+d$，要满足 $17+d$ 是 9 的倍数，选项 (A) 中 $d=1$ 是唯一可能的答案。

2.3.4　Divisibility Rule For 7, 11, and 13 可以被 7，11 和 13 整除的规则

1. If you double the last digit and subtract it from the rest of the number and the answer is divisible by 7, the number is divisible by 7. Repeat this process on the new result if needed. (Take 525 as an example. Since $52-5\times2=42$ and 42 is divisible by 7, we know 525 is divisible by 7.)

如果你将末位数字加倍，并从剩下的数字中减去它，若结果可以被 7 整除，则这个数可以被 7 整除。若数字仍比较大，你可以对结果重复应用此规则。(以 525 为例。由于 $52-5\times2=42$ 且 42 能被 7 整除，所以 525 能被 7 整除。)

2. To find out if a number is divisible by 11, add every other digit, and call that sum “x”. Add together the remaining digits, and call that sum “y”. Take the positive difference of x and y. If the difference is zero or a multiple of eleven, then the original number is a multiple of eleven. (Take 685421 as an example. The sum of every other digit could be $1+4+8=$

13, and the sum of the rest digits is $2+5+6=13$. The absolute value of their difference is $|13-13|=0$. So, 685421 is divisible by 11.)

要判断一个数是否可以被 11 整除，将每隔一位的数字相加，得出和为“x”。将剩余位数的数字相加，得出和为“y”。取 x 和 y 的差的绝对值，如果为零或 11 的倍数，则原数可以被 11 整除。(以 685421 为例。每隔一位数字相加得 $1+4+8=13$，剩余位数的数字相加得 $2+5+6=13$，它们的差的绝对值为 $|13-13|=0$，那么 685421 能被 11 整除。)

3. Delete the last digit from the number, and then subtract 9 times the deleted digit from the remaining number. If what is left is divisible by 13, then so is the original number. Repeat the rule if necessary. (Take 6448 as an example. First we remove the last digit 8 and get 644. Then we subtract it by $8\times9=72$, and have $644-72=572$. It is still hard to tell if 572 is divisible by 13, so we apply the rule again. We get $57-2\times9=39$. Since 39 is divisible by 13, we can tell 6448 is divisible by 13.)

从一个数中删除末位数字，然后从剩余数字中减去被删除的数字的 9 倍。若结果可以被 13 整除，则原数也可以被 13 整除。若得到的数仍比较大，可以重复应用此规则。(以 6448 为例。首先删除末位数字 8 变成 644，然后减去 $8\times9=72$，$644-72=572$。仍然很难判断 572 是否能被 13 整除，因此我们再次应用该法则，得到 $57-2\times9=39$。因为 39 能被 13 整除，所以 6448 能被 13 整除。)

4. If the positive difference of the last three digits and the rest of the digits is divisible by 7, 11, or 13, then the number is divisibly by 7, 11, or 13, respectively. (Take 645603 as an example. The last three digits are 603, and the rest digits are 645. Since $|645-603|=42$ and it is divisible by 7, we know 645603 is divisible by 7.)

如果末三位数字和其余数字的绝对差值可以被 7，11 或 13 整除，则这个数可以被 7，11 或 13 整除。(以 645603 为例。末三位数字为 603，剩余数字为 645。差的绝对值为 $|645-603|=42$，结果能被 7 整除，那么原数能被 7 整除。)

Try to find the number 49896, whether it can be divisible by 2, 4, 8, 5, 3, 9, 7, 11, 13 or not?

试着判断 49896 是否可以被 2、4、8、5、3、9、7、11 或 13 整除?

2.3.5 Divisibility Rule For 6, 10, 12, 14, 15, 18, 24, and 36 可以被 6，10，12，14，15，18，24 和 36 整除的规则

- A number is divisible by 6 if the number is divisible by both 2 and 3.

- A number is divisible by 10 if the number is divisible by both 2 and 5.
- A number is divisible by 12 if the number is divisible by both 3 and 4.
- A number is divisible by 14 if the number is divisible by both 2 and 7.
- A number is divisible by 15 if the number is divisible by both 3 and 5.
- A number is divisible by 18 if the number is divisible by both 2 and 9.
- A number is divisible by 24 if the number is divisible by both 3 and 8.
- A number is divisible by 36 if the number is divisible by both 4 and 9.
- 一个数如果可以同时被 2 和 3 整除，则它可以被 6 整除。
- 一个数如果可以同时被 2 和 5 整除，则它可以被 10 整除。
- 一个数如果可以同时被 3 和 4 整除，则它可以被 12 整除。
- 一个数如果可以同时被 2 和 7 整除，则它可以被 14 整除。
- 一个数如果可以同时被 3 和 5 整除，则它可以被 15 整除。
- 一个数如果可以同时被 2 和 9 整除，则它可以被 18 整除。
- 一个数如果可以同时被 3 和 8 整除，则它可以被 24 整除。
- 一个数如果可以同时被 4 和 9 整除，则它可以被 36 整除。

NOTE: If a number is divisible by two numbers that are relatively prime, then it is divisible by the product of those two numbers. "Relatively prime" means two numbers have no common factor other than 1. For example, 3 and 4 are relatively prime.

If a number is divisible by n numbers that are relatively prime, then it is divisible by the product of those n numbers.

注意：如果一个数可以被两个互质的数整除，那么它也可以被这两个数的乘积整除。"互质"的意思是两个数除了 1 之外没有其他共同因子。例如，3 和 4 是互质的。

如果一个数可以被 n 个互质的数整除，那么它也可以被这 n 个数的乘积整除。

Example 7

Find distinct digits A and B such that A32B is as large as possible and divisible by 36. Name the number.

例题 7

找到不同的数字 A 和 B，使得 A32B 尽可能大且能被 36 整除。请列出这个数字。

(A) 9322 (B) 9321 (C) 9320 (D) 9324 (E) 9328

Solution

A32B is divisible by 36, so it is also divisible by 9 and 4. The sum of digits is A+3+2+B=A+B+5. Since A and B are distinct and 2B needs to be divisible by 4, we know B=0, 4 or 8. Since A32B is divisible by 9, we know A+B+5 is divisible by 9. If B=0, we have A=4; if B=4, we have A=9; if B=8, we have A=5. To get the greatest value of A32B, we want to maximize A, so A = 9 and B=4. Therefore, A23B=(D)9324.

解 析

A32B 可以被 36 整除，因此它也可以被 9 和 4 整除。总和为 A+3+2+B=A+B+5。由于 A 和 B 是不同的，末尾两个数字 2B 必须能被 4 整除，所以 B=0，4 或 8。因为 A32B 必须能被 9 整除，所以 A+B+5 能被 9 整除。当 B=0 时，可得 A=4；当 B=4 时，可得 A=9；当 B=8 时，可得 A=5。为使 A32B 值最大，我们使 A 的值最大。所以 A=9 且 B=4。因此，A32B = 9324，答案是 (D)。

Example 8

A number is called flippy if its digits alternate between two distinct digits. For example, 2525 and 46464 are flippy, but 2442 and 789789 are not. How many five-digit flippy numbers are divisible by 15?

例题 8

如果一个数的每一位数字在两个不同的数字之间交替变化，则称该数字为“flippy”。例如，2525 和 46464 是 flippy，但 2442 和 789789 不是 flippy。问共有多少个五位 flippy 可以被 15 整除？

(A) 3 (B) 4 (C) 5 (D) 6 (E) 8

Solution

Since the flippy number is divisible by 15, it must be divisible by 3 and 5. By the divisibility of 5, we know the last digit is 0 or 5. Since the number is a 5-digit flippy number, the last digit is the same as the first digit, resulting the last digit not being 0. Thus, the number is of the form $5x5x5$. We can find that one of the solutions to x is 0, since $5+5+5$ is equal to 15, which is divisible by 3. After trying every one-digit number, you'll notice that x must be a multiple of 3, meaning that $x = 0$, 3, 6, or 9. Therefore, 50505, 53535, 56565, and 59595 are the solutions to this question. So the answer is (B).

解 析

因为这个数能被 15 整除，所以必须同时被 3 和 5 整除，即最后一位数为 0 或 5，但根据数位重复的性质，0 不可以放在首位，所以最后一位数必须是 5，即这个数的形式为 $5x5x5$。因为 $5+5+5=15$ 已经是 3 的倍数，因此其中一个解是 $x=0$。将 0-9 所有数代入后，得 $x=0$、3、6 或 9。所以，50505、53535、56565 和 59595，一共有 4 个数满足题意，答案是 (B)。

Example 9

In Theresa's first 8 basketball games, she scored 5, 1, 2, 3, 7, 8, 6 and 5 points. In her ninth game, she scored fewer than 10 points and her points per-game average for the nine games was an integer. Similarly in her tenth game, she scored fewer than 10 points and her points-per-game average for the 10 games was also an integer. What is the sum of the number of points she scored in the ninth and tenth games?

例题 9

在特丽萨的前 8 场篮球比赛中，她的得分分别是 5，1，2，3，7，8，6 和 5 分。在她第 9 场比赛中，她的得分少于 10 分且前 9 场的平均得分是一个整数。同样，在第 10 场比赛中，她得分也少于 10 分且前 10 场比赛的平均得分也是一个整数。她在第 9 场和第 10 场比赛中得分的和是多少？

(A) 8　　(B) 12　　(C) 13　　(D)15　　(E) 16

Solution

The total number of points from the first 8 games is $7+4+3+6+8+3+1+5=37$. We have to make this a multiple of 9 by scoring less than 10 points. The closest multiple of 9 is 45, and $45-37=8$. Then we add a number to get a multiple of 10. The next multiple is 50, so the added score is 5. Adding these two scores together you get $8+5=13$. The answer is (C).

解 析

前 8 场比赛的总得分是 $7+4+3+6+8+3+1+5=37$。第 9 场比赛得分少于 10 分，且 9 场总得分是 9 的倍数。最接近的 9 的倍数是 45，$45-37=8$。同理，前 10 场比赛得分总和是 10 的倍数，即 45 再加一个数，是 10 的倍数。下一个 10 的倍数是 50，所以加起来的分数为 5 分。将它们相加，得到 $8+5=13$。答案是 (C)。

Example 10

How many three-digit numbers are divisible by 17?

例题 10

有多少个三位数可以被 17 整除?

(A) 5　　(B) 52　　(C) 53　　(D) 58　　(E) 67

Solution

Let k be a positive integer, then $17k$ is a multiple of 17. Suppose $17k$ to be a three-digit number. Since the smallest 3-digit number is 100, we have $17k>100$ and $k>5.8$. Since the greatest 3-digit number is 999, we have $17k<999$ and $k<58.7$. Thus, the number k can range from 6 to 58 and there are (C)53 such numbers.

解 析

设 k 为任意正整数，则 $17k$ 是 17 的倍数。假设 $17k$ 是一个三位数。最小的三位数为 100，因此 $17k>100$，得 $k>5.8$。最大的三位数为 999，因此 $17k<999$ 并得 $k<58.7$。因此，k 的取值范围是从 6 到 58（包含 6 和 58），所以有 53 个三位数。故答案是 (C)。

PRACTICE 练习

Problem 1

The sum of all multiples of 3 from 10 to 50 is S. Is S even or odd?

问题 1

从 10 到 50 的所有 3 的倍数的和是 S。问 S 是偶数还是奇数?

(A) even 偶数　　　　(B) odd 奇数

Problem 2

If x and y are integers and xy^2+x^2 is odd, which of the following statements must be true?

问题 2

如果 x 和 y 是整数且 xy^2+x^2 是奇数，以下哪个选项一定正确?

Ⅰ. x^2 is odd.　x^2 是奇数。

Ⅱ. x is odd.　x 是奇数。

Ⅲ. y^2 is odd. y^2 是奇数。

(A) Ⅰ only　　　　只有Ⅰ

(B) Ⅲ only　　　　只有Ⅲ

(C) Ⅰ and Ⅱ　　　　Ⅰ和Ⅱ

(D) Ⅰ and Ⅲ　　　　Ⅰ和Ⅲ

(E) Ⅱ and Ⅲ　　　　Ⅱ和Ⅲ

Problem 3

When Yunji added all the integers from 1 to 9, she mistakenly left out a number. Her incorrect sum turned out to be a prime number. What is the sum of all the possible numbers she might have missed?

问题 3

云姬将 1～9 的所有整数相加，她不小心漏了一个数，得到的总和是一个质数。她可能漏掉的数的总和是多少?

(A) 3　　(B) 4　　(C) 8　　(D) 14　　(E) 16

Problem 4

Suppose a, b, and c are nonzero real numbers, and $a+b+c=0$. What are the possible value(s) for $\frac{|2a|}{a}+\frac{|2b|}{b}+\frac{|2c|}{c}+\frac{|2abc|}{abc}$?

问题 4

假设 a，b 和 c 是非零实数，且 $a+b+c=0$。那么 $\frac{|2a|}{a}+\frac{|2b|}{b}+\frac{|2c|}{c}+\frac{|2abc|}{abc}$ 的值可能是多少？

(A) 0 0

(B) 1 and –1 1 和 –1

(C) 2 and –2 2 和 –2

(D) 0, 2, and –2 0，2 和 –2

(E) 0, 1, and –1 0，1 和 –1

Problem 5

The sum of 25 consecutive even integers is 9000. Which is the largest of these 25 consecutive integers?

问题 5

25 个连续偶数的和是 9000。这 25 个连续偶数中最大的一个是多少？

(A) 384 (B) 488 (C) 512 (D) 614 (E) 628

Problem 6

What is the smallest possible average of four distinct positive odd integers?

问题 6

四个不同的正奇数的最小可能平均值是多少？

(A) 3 (B) 4 (C) 5 (D) 6 (E) 7

Problem 7

The digits 1, 2, 3, 4, and 5 are each used once to write a five-digit number *ABCDE*. The three-digit number *ABC* is divisible by 4, the three-digit number *BCD* is divisible by 5, and the three-digit number *CDE* is divisible by 3. What is *A*?

问题 7

数字 1，2，3，4 和 5 各使用一次来写一个五位数 *ABCDE*。其中三位数 *ABC* 必须能被 4 整除，三位数 *BCD* 必须能被 5 整除，三位数 *CDE* 必须能被 3 整除。请问 *A* 是多少？

(A) 1 (B) 2 (C) 3 (D) 4 (E) 5

Problem 8

Minh enters the numbers 1 through 100 into the cells of a 10×10 grid in some order. She calculates the product of the numbers in each row and column. What is the least number of rows and columns that could have a product divisible by 5?

问题 8

明将 1 ～ 100 这 100 个数以某种顺序填入一个 10×10 的网格中。她计算每一行和每一列的乘积。最少有多少行和列的乘积能够被 5 整除？

(A) 5　　(B) 6　　(C) 9　　(D) 12　　(E) 18

Problem 9

Find the value of x such that the four-digit number $x65x$ is divisible by 18.

问题 9

找出使四位数 $x65x$ 能被 18 整除的 x 的值。

(A) 1　　(B) 2　　(C) 4　　(D) 5　　(E) 8

Problem 10

Let Z be a 6-digit positive integer, such as 123123, whose first three digits are the same as its last three digits taken in the same order. Which of the following numbers might not be a factor of Z?

问题 10

设 Z 是一个 6 位的正整数，比如 123123，其前面三位数与后三位数相同且顺序一致。以下哪个数字不一定是 Z 的因数？

(A) 11　　(B) 17　　(C) 1001　　(D) 3　　(E) 7

Problem 11

Eleven members of the Middle School Math Club each paid the same integer amount for a guest speaker to talk about problem solving at their math club meeting. In all, they paid their guest speaker \$3A2. What is the missing digit A of this 3-digit number?

问题 11

中学数学俱乐部的 11 名成员每人支付了相同的整数金额来邀请嘉宾进行数学讲座。在总共支付给嘉宾的金额 3A2 美元中，缺失的数字 A 是多少？

(A) 1　　(B) 2　　(C) 3　　(D) 4　　(E) 5

Problem 12

Anna wants to arrange her model cars in rows with exactly 6 cars in each row. She now has 21 model cars. What is the smallest number of additional cars she must buy in order to be able to arrange all her cars this way?

问题 12

安娜想把她的模型车按每行 6 辆车的方式排列。她现在有 21 辆模型车。为了能按这种方式排列她所有的车，她最少还需要买多少辆模型车？

(A) 1 (B) 2 (C) 3 (D) 4 (E) 5

Problem 13

The sequence 1, 1, 2, 4, 7, 13, 24, ⋯ is formed like this: any term is the sum of the three terms before it starts from the fourth term. Is the 300th term even or odd in the sequence?

问题 13

数列 1, 1, 2, 4, 7, 13, 24, ⋯ 的规律是：从第四项开始每一项都是前面三项的和。请问数列中的第 300 项是偶数还是奇数？

(A) Even 偶数 (B) Odd 奇数

Problem 14

The sequence 1, 1, 2, 3, 5, 8, 13, 21, ⋯ is formed like this: any term is the sum of the two terms before it, starting from the third term. How many even numbers are there in the first 900 terms of the sequence?

问题 14

数列 1，1，2，3，5，8，13，21，⋯ 的规律是：从第三项开始，每一项都是前面两项的和。在前 900 项中，有多少个偶数？

(A) 300 (B) 400 (C) 500 (D) 600 (E) 650

Problem 15

Mr. Mathis and his student Peter worked together to solve math problems last week. When each person solved a problem, that person put a marble in his own box. Mr. Mathis's problem solving speed was double of his student. At the end of the problem solving session, Peter had 2 boxes of marbles and Mr. Mathis had 4 boxes of marbles. Each box is labeled with the number of marbles inside it. The numbers are 60, 65, 57, 52, 68 and 73, respectively.

Which two boxes do not belong to Mr. Mathis?

问题 15

马西斯先生和他的学生彼得上周一起解数学题。每人每解决一个问题，都会在自己的盒子里放一个弹珠。马西斯先生的解题速度是他学生的 2 倍。解题结束时，彼得有两个盒子，马西斯先生有四个盒子。每个盒子上标有其中的弹珠数量，分别是 60，65，57，52，68 和 73。请问哪两个盒子不属于马西斯先生？

(A) 60 and 65　　60 和 65

(B) 57 and 65　　57 和 65

(C) 57 and 73　　57 和 73

(D) 52 and 68　　52 和 68

(E) 60 and 68　　60 和 68

Chapter 3 Base Representation and Digits

第 3 章 基数和位数

3.1 Basic Knowledge 基础内容

3.1.1 Base 10 number representations 十进制表达

$M=\overline{a_n a_{n-1}\cdots a_1 a_0}=10^n a_n+10^{n-1}a_{n-1}+\cdots+10^1 a_1+10^0 a_0$,

$M=\overline{a_n a_{n-1}\cdots a_1 a_0}=10\times\overline{a_n a_{n-1}\cdots a_1}+a_0$,

$M=\overline{a_n a_{n-1}\cdots a_1 a_0}=100\times\overline{a_n a_{n-1}\cdots a_2}+\overline{a_1 a_0}$,

$M=\overline{a_n a_{n-1}\cdots a_1 a_0}=10^k\times\overline{a_n a_{n-1}\cdots a_k}+\overline{a_{k-1}a_{k-2}\cdots a_1 a_0}$,

$M=\overline{2345}=2\times10^3+3\times10^2+4\times10^1+5\times10^0=2000+300+40+5$,

$M=\overline{2345}=10\times\overline{234}+5$,

$M=\overline{2345}=100\times\overline{23}+45$,

$M=\overline{2345}=1000\times\overline{2}+345$,

$$M=\overline{a_n a_{n-1}\cdots a_1 a_0.\,b_1 b_2\cdots b_m}$$
$$=10^n a_n+10^{n-1}a_{n-1}+\cdots+10^1 a_1+10^0 a_0+10^{-1}b_1+10^{-2}b_2+\cdots+10^{-m}b_m,$$

$$M=\overline{23.45}=2\times10^1+3\times10^0+4\times10^{-1}+5\times10^{-2}$$
$$=20+3+0.4+0.05.$$

3.1.2 Patterns of the last digit of a^n a^n 个位数的规律

The last digits of a^n have patterns shown in the table below.

在下面的表格中，显示了数字 a^n 个位数的规律。

n	1	2	3	4	Period 周期
2^n	**2**	**4**	**8**	**6**	4
3^n	**3**	**9**	**7**	**1**	4
4^n	**4**	**6**			2
5^n	**5**				1
6^n	**6**				1
7^n	**7**	**9**	**3**	**1**	4
8^n	**8**	**4**	**2**	**6**	4
9^n	**9**	**1**			2

For example, when $a=2$,

比如，当 $a=2$ 时，

$$2^1=2,$$
$$2^2=4,$$
$$2^3=8,$$
$$2^4=16,$$
$$2^5=32,$$
$$2^6=64,$$
$$2^7=128,$$
$$2^8=256,$$
$$\cdots$$

The last digits of 2^n demonstrate a pattern: 2, 4, 8, 6, 2, 4, 8, 6, $\cdots$.

2^n 的个位数有如下规律：2，4，8，6，2，4，8，6，$\cdots$。

Example 1

A 2-digit number is such that the sum of the digits multiplied by two is equal to the number. What is the unit's digit of the number?

例题 1

一个两位数满足各数位数字的和的两倍等于该两位数，那么这个两位数的个位数字是多少？

(A) 2　　(B) 4　　(C) 5　　(D) 6　　(E) 8

Solution

We can think of the number as $10a+b$, where a is the tens digit and b is the units digit. Since the sum of the digits is $a+b$, we can have $2(a+b)=10a+b$. Simplify this to $8a=b$, and get that b is a multiple of 8. Since b is the units digit, it could be 0 or 8. If $b=0$, then $a=0$, contradicting that a is the leading digit. Therefore, the units digit b is (E) 8.

解 析

根据题意，可以将这个数表示为 $10a+b$，其中 a 是十位数，b 是个位数。则数位之和为 $a+b$，因此 $2(a+b)=10a+b$。将其简化为 $8a=b$，因此，b 是 8 的倍数。又因为 b 只能是一个一位数，所以 b 是 0 或者 8。如果 $b=0$，则 $a=0$，和 a 是首位数矛盾。因此个位数 b 只能是 8，答案是 (E)。

Example 2

When Clara totaled her scores, she inadvertently reversed the units digit and the tens digit of one score. By which of the following might her incorrect sum have differed from the correct one?

例题 2

当克拉汇总她的得分时，她不小心将得分的个位数和十位数颠倒了。请问错误的与正确的数加一起可能是多少？

(A) 34 (B) 35 (C) 36 (D) 44 (E) 45

Solution

Let the two digits be a and b. The correct score was $10a+b$. Clara misinterpreted it as $10b+a$. The sum between the two is $11a+11b$, which factors into $11(a+b)$. Therefore, since the sum is a multiple of 11, the only answer choice that is a multiple of 11 is (D) 44.

解 析

设正确得分的十位数字为 a，个位数字为 b，则正确的得分是 $10a+b$。克拉颠倒的得分为 $10b+a$，两者的和是 $11a+11b$，提取公因数后可得 $11(a+b)$。由于该数是 11 的倍数，唯一是 11 的倍数的选项是 (D)。

Example 3

An ATM password at a bank is composed of six digits from 0 to 9, with repeated digits allowable. If no password may begin with the sequence 1, 1, 0, then how many passwords are possible?

例题 3

一家银行的 ATM 密码由 6 个数字（0 ～ 9）组成，可以重复。如果密码不能以序列 1，1，0 开头，那么可能的密码数量是多少？

(A) 3000　(B) 729000　(C) 900000　(D) 999000　(E) 999999

Solution

For the first three digits, there are $10^3-1=999$ combinations since 110 is not allowed. For the last three digit, any of the 10 numbers are allowed. Therefore, the total number of passwords is: $999\times10^3=(\text{D})999000$.

解　析

前面三个数字不能以 1，1，0 开头，因此有 $10^3-1=999$ 种组合。对于后面三个数字，任何 10 个数字都可以使用。因此，可能的密码总数是 $999\times10^3=999000$，故选 (D)。

Example 4

Let the letters $\underline{C}\underline{A}\underline{R}\underline{P}\underline{E}\underline{T}$ represent distinct digits. Suppose $\underline{C}\underline{A}\underline{R}\underline{C}\underline{A}\underline{R}$ is the greatest number that satisfies the equation

$$6\times\underline{C}\underline{A}\underline{R}\underline{C}\underline{A}\underline{R}=\underline{P}\underline{E}\underline{T}\underline{P}\underline{E}\underline{T}$$

What is the value of $\underline{C}\underline{A}\underline{R}+\underline{P}\underline{E}\underline{T}$?

例题 4

设 $\underline{C}\underline{A}\underline{R}\underline{P}\underline{E}\underline{T}$ 中每个字母表示不同的数字。假设 $\underline{C}\underline{A}\underline{R}\underline{C}\underline{A}\underline{R}$ 是满足方程 $6\times\underline{C}\underline{A}\underline{R}\underline{C}\underline{A}\underline{R}=\underline{P}\underline{E}\underline{T}\underline{P}\underline{E}\underline{T}$ 的最大数字。那么 $\underline{C}\underline{A}\underline{R}+\underline{P}\underline{E}\underline{T}$ 的值是多少？

(A) 1141　(B) 1148　(C) 1155　(D) 1162　(E) 1169

Solution

Notice that $\underline{C}\underline{A}\underline{R}\underline{C}\underline{A}\underline{R} = 1000(\underline{C}\underline{A}\underline{R}) + \underline{C}\underline{A}\underline{R}$.

Likewise, $\underline{P}\underline{E}\underline{T}\underline{P}\underline{E}\underline{T} = 1000(\underline{P}\underline{E}\underline{T}) + \underline{P}\underline{E}\underline{T}$.

Therefore, we have the following equation:

$$6 \times 1001(\underline{C}\underline{A}\underline{R}) = 1001(\underline{P}\underline{E}\underline{T}) .$$

Simplify the equation and get $6 \times (\underline{C}\underline{A}\underline{R}) = \underline{P}\underline{E}\underline{T}$.

Now we verify each choice using this equation. Since $\underline{C}\underline{A}\underline{R} + \underline{P}\underline{E}\underline{T} = 7(\underline{C}\underline{A}\underline{R})$, we can get the value of $(\underline{C}\underline{A}\underline{R})$ by dividing the choices using 7, i.e.,

A. $\underline{CAR} = 1141 \div 7 = 163$, $\underline{PET} = 6 \times \underline{CAR} = 978$

B. $\underline{CAR} = 1148 \div 7 = 164$, $\underline{PET} = 6 \times \underline{CAR} = 984$ (digit 4 is repeated)

C. $\underline{CAR} = 1155 \div 7 = 165$, $\underline{PET} = 6 \times \underline{CAR} = 990$ (digit 9 is repeated)

D. $\underline{CAR} = 1162 \div 7 = 166$ (digit 6 is repeated)

E. $\underline{CAR} = 1169 \div 7 = 167$, $\underline{PET} = 6 \times \underline{CAR} = 1002$ (a 4-digit number)

Only (A) gives a qualified answer.

So, the correct answer is (A).

解　析

因为 $\underline{C}\underline{A}\underline{R}\underline{C}\underline{A}\underline{R} = 1000(\underline{C}\underline{A}\underline{R}) + \underline{C}\underline{A}\underline{R}$，而且 $\underline{P}\underline{E}\underline{T}\underline{P}\underline{E}\underline{T} = 1000(\underline{P}\underline{E}\underline{T}) + \underline{P}\underline{E}\underline{T}$。

可得以下方程：$6 \times 1001(\underline{C}\underline{A}\underline{R}) = 1001(\underline{P}\underline{E}\underline{T})$。简化得 $6 \times (\underline{C}\underline{A}\underline{R}) = \underline{P}\underline{E}\underline{T}$。

将每个选项代入方程进行验算。$\underline{C}\underline{A}\underline{R} + \underline{P}\underline{E}\underline{T} = 7(\underline{C}\underline{A}\underline{R})$，将每个答案除以 7 得到 $\underline{C}\underline{A}\underline{R}$ 的值，具体如下：

A. $\underline{CAR} = 1141 \div 7 = 163$，$\underline{PET} = 6 \times \underline{CAR} = 978$

B. $\underline{CAR} = 1148 \div 7 = 164$，$\underline{PET} = 6 \times \underline{CAR} = 984$（数字 4 重复了）

C. $\underline{CAR} = 1155 \div 7 = 165$，$\underline{PET} = 6 \times \underline{CAR} = 990$（数字 9 重复了）

D. $\underline{CAR} = 1162 \div 7 = 166$（6 重复了）

E. $\underline{CAR} = 1169 \div 7 = 167$，$\underline{PET} = 6 \times \underline{CAR} = 1002$（四位数，不符）

由于每个字母代表不一样的数字，只有选项 (A) 满足条件。

因此，正确答案是 (A)。

3.2 Remainder 余数

3.2.1 Definition 定义

For $b \geqslant 2$, we write $b|(a-r)$ if $a-r$ is divisible by b.

The following expressions mean the same thing: (q is the quotient and has an integer value)

$$b|(a-r)$$
$$a-r=qb$$
$$a=qb+r$$
$$\frac{a-r}{b}=q$$

对于 $b \geqslant 2$，我们用 $b|(a-r)$ 表示 $a-r$ 可以被 b 整除。

以下表达式的意思相同：（q 表示商，且为整数值）

$$b|(a-r)$$
$$a-r=qb$$
$$a=qb+r$$
$$\frac{a-r}{b}=q$$

3.2.2 Theorem 1 定理 1

There exists a unique pair (q, r) such that:

$$a=qb+r$$

or

$$\frac{a}{b}=q+\frac{r}{b}$$

where a and b are integers, $b > 0$, q is the quotient, and r is the remainder with $0 \leqslant r < b$.

If $r = 0$, $\frac{a}{b}=q+\frac{r}{b}$ which can be written as $a=qb$. This is equivalent to saying that a is divisible by b, $b|a$, and b divides a.

存在一对唯一的 (q, r)，使得

$$a=qb+r$$

或

$$\frac{a}{b}=q+\frac{r}{b}$$

其中 a 和 b 是整数，且 $b>0$，q 是商，r 是余数且满足 $0 \leqslant r<b$。

如果 $r=0$，$\frac{a}{b}=q+\frac{r}{b}$ 可以写作 $a=qb$。这等同于说 a 可以被 b 整除，即 $b|a$，并且 b 是 a 的因数。

3.2.3 Theorem 2 (Residue Class) 定理 2（剩余类）

For any given positive integer m, when it is divided by n, the remainder must be one the following: 0, 1, 2, ⋯ , $n-1$. All integers can be classified into a unique class according to their remainder when divided by n. A complete set of residue classes has n classes.

Any integer can be classified into two residue classes when divided by 2:

$2k$, (remainder is 0)

$2k+1$. (remainder is 1)

Here k is an integer.

Any integer can be classified into 4 classes when divided by 4:

$4k$, (remainder is 0)

$4k+1$, (remainder is 1)

$4k+2$, (remainder is 2)

$4k+3$. (remainder is 3)

Here k is an integer.

对于任何给定的正整数 m，当它被一个数 n 除时，余数必定是以下之一：0，1，2，⋯，$n-1$。所有整数可以根据它们被 n 除时的余数分类为唯一的类别。一个完整的剩余类集合包含 n 个类别。

任何整数在被 2 除时可以分类为两个剩余类：

$2k$，（余数为 0）

$2k+1$，（余数为 1）

其中 k 是一个整数。

任何整数在被 4 除时可以分类为四个剩余类：

$4k$，（余数为 0）

$4k+1$，（余数为 1）

$4k+2$，（余数为 2）

$4k+3$，（余数为 3）

其中 k 是一个整数。

Example 5

The number N is a two-digit number.

When N is divided by 9, the remainder is 2.

When N is divided by 10, the remainder is 5.

What is the remainder when N is divided by 13?

例题 5

数字 N 是一个两位数。

当 N 被 9 除时，余数是 2。

当 N 被 10 除时，余数是 5。

当 N 被 13 除时，余数是多少?

(A) 0　　(B) 3　　(C) 6　　(D) 9　　(E) 10

Solution

Since the remainder of N when divided by 10 is 5, we know that the units digit must be 5. Since the remainder of N is 2 when divided by 9, N is the sum of a multiple of 9 and 2, and the multiple of 9 must end in 3. We now check the multiples of 9:

$$9(1)=9,\ 9(2)=18,\ 9(3)=27,\ 9(4)=36,$$

$$9(5)=45,\ 9(6)=54,\ 9(7)=63,\ 9(8)=72.$$

63 is the only qualified 2-digit multiple, so N is $9\times7+2=65$. Since $65\div13=5$, the remainder is (A) 0.

解　析

N被10整除时余数为5，可得该数的个位数字必须是5。又因N被9除时有余数2，所以N是一个9的倍数与2的和，且这个9的倍数的个位为3。我们来看9的倍数:

$$9(1)=9,\ 9(2)=18,\ 9(3)=27,\ 9(4)=36,$$

$$9(5)=45,\ 9(6)=54,\ 9(7)=63,\ 9(8)=72。$$

发现只有63满足条件，因此N为$9\times7+2=65$。因为$65\div13=5$，所以余数为0，故选(A)。

Example 6

The 5-digit number 2024A is divisible by 9. What is the remainder when this number is divided by 5?

例题 6

5 位数 2024A 能被 9 整除。这个数字被 5 除时的余数是多少?

(A) 0 (B) 1 (C) 2 (D) 3 (E) 4

Solution

We use the property that the digits of a number must sum to a multiple of 9 if it is divisible by 9. This means $2+0+2+4+A$ must be divisible by 9. The only possible value for A then must be 1. Since we are looking for the remainder when divided by 5, then we can ignore the first four digits and only focus on the unit digit. The remainder when 20241 is divided by 5 is (B) 1.

解 析

根据性质如果该数可以被 9 整除，那么各位数字之和必须是 9 的倍数。因此 $2+0+2+4+A=8+A$ 必须能被 9 整除。A 唯一可能的值是 1。因为我们关注被 5 整除时的余数，所以可以忽略前 4 位而只关心个位数。20241 被 5 除时的余数是 1，故选 (B)。

Example 7

A box contains gold coins. If the coins are equally divided among seven people, four coins are left over. If the coins are equally divided among six people, three coins are left over. If the box holds the smallest number of coins that meets these two conditions, how many coins are left when equally divided among nine people?

例题 7

一个箱子里装着金币。如果将金币平均分给七个人，还剩下四枚。如果将金币平均分给六个人，还剩下三枚。如果箱子里装着满足这两个条件的最少数量的金币，那么当金币平均分给九个人时，还剩下多少枚金币？

(A) 0　(B) 1　(C) 3　(D) 5　(E) 7

Solution

The counting numbers that leave a remainder of 4 when divided by 7 are 11, 18, 25, 32, 39, 46, 53, 60,⋯.

The counting numbers that leave a remainder of 3 when divided by 6 are 9, 15, 21, 27, 33, 39, 45, 51,⋯.

39 is the smallest possible number of coins that meets both conditions. When 39 is divided by 9, the remainder is 3. Therefore, there are (C) 3 coins left.

解　析

当被 7 除时余数为 4 的数是：11，18，25，32，39，46，53，60，⋯。

当被 6 除时余数为 3 的数是：9，15，21，27，33，39，45，51，⋯。

因此 39 是满足这两个条件的最小可能的金币数量。而当 39 被 9 除时余数为 3。所以当它们被九个人分的时候会剩下 3 个金币，故选 (C)。

3.2.4　Theorem 3 定理 3

The largest number which divides any two given numbers leaving the same remainder equals the difference of the two numbers.

能整除任意两个给定数字并得到相同余数的最大数等于这两个给定数字的差。

Example 8

A number N divides each of 17 and 30 with the same remainder in each case. What is the largest value of N?

例题 8

17 和 30 被 N 除后有相同的余数。N 的最大值是多少？

(A) 11　(B) 12　(C) 13　(D) 15　(E) 16

Solution

By the theorem 3, the largest $N = 30 - 17 = 13$. The answer is (C).

解 析

根据定理 3，N 的最大值为 $N = 30 - 17 = 13$。故答案为 (C)。

3.2.5 Theorem 4 定理 4

If $a|b$, $a|c$, then $a|(jb + kc)$, and j, k are any integers.

This means if a can divide b and c, then a can divide any linear combination $jb + kc$ of b and c, where j and k are aribitrary integers.

如果 $a|b$ 且 $a|c$，则 $a|(jb + kc)$，其中 j 和 k 是任意整数。

也就是说，如果 a 能整除 b 且能整除 c，那么 a 能整除 b 和 c 的线性组合 $jb + kc$，其中 j 和 k 是任意整数。

Example 9

A number N divides each of 17 with the remainder of r and 30 with the remainder $2r$ respectively. What is the largest value of N?

例题 9

一个整数 N 去除 17 余数为 r，同一个整数 N 去除 30 余数为 $2r$。求 N 的最大值是多少？

(A) 2 (B) 3 (C) 4 (D) 5 (E) 6

Solution

We know that N divides both $17 - r$ and $30 - 2r$. By Theorem 4, N also divides any linear combination of them. That is, N divides $2(17 - r) - (30 - 2r) = 4$. Thus, N is a factor of 4, so the greatest N is (C)4.

Note that we can also use Theorem 3 to solve it. Since the remainder of dividing by 17 is r, we know the remainder of dividing by 34 is $2r$, which is the same as dividing by 30. By Theorem 3, the greatest N is 4.

解析

由题可知 N 可以整除 $17-r$ 和 $30-2r$。由定理 4 可知，N 同样也可以整除这两个数的线性组合式。即 N 可以整除 $2(17-r)-(30-2r)$，展开得 $2(17-r)-(30-2r)=4$，所以 N 是 4 的一个因子，所以 N 最大值为 4。故答案为 (C)。

注意，我们也可以利用定理 3 来解决这个问题。由于除以 17 的余数是 r，我们知道除以 34 的余数是 $2r$，这与除以 30 的余数相同。根据定理 3，最大的 N 值是 4。

PRACTICE 练习

Problem 1

What is the ones digit of $333333-33333-3333-333-33-3$?

问题 1

表达式 $333333-33333-3333-333-33-3$ 的个位数是什么?

(A) 1　　(B) 3　　(C) 4　　(D) 7　　(E) 8

Problem 2

The hundreds digit of a three-digit number is 4 less than the units digit. The digits of the three-digit number are reversed, and use the result subtract the original three-digit number. What is the units digit of the result?

问题 2

一个三位数的百位数字比个位数字小 4。将这个三位数的数字顺序反转，用得到的结果减去原来的三位数。最终结果的个位数字是多少?

(A) 0　　(B) 2　　(C) 4　　(D) 6　　(E) 8

Problem 3

The letters A, B, C and D represent digits. If

$$\begin{array}{r} A\ B \\ +\ C\ A \\ \hline 1\ 0\ A \end{array} \quad \text{and} \quad \begin{array}{r} A\ B \\ -\ C\ A \\ \hline 1\ C \end{array}$$

what digit does C represent?

问题 3

字母 A，B，C 和 D 代表数字。如果满足

$$\begin{array}{rcc} & A & B \\ + & C & A \\ \hline 1 & 0 & A \end{array} \text{ 和 } \begin{array}{rcc} & A & B \\ - & C & A \\ \hline & 1 & C \end{array}$$

那么字母 C 代表哪个数字？

(A) 2 (B) 3 (C) 4 (D) 5 (E) 6

Problem 4

The members of a band are arranged in a rectangular formation. When they are arranged in 8 rows, there are 2 positions unoccupied in the formation. When they are arranged in 9 rows, there are 3 positions unoccupied. How many members are in the band if the membership is between 100 and 200?

问题 4

一个乐队的成员以矩形阵形排列。当他们排列成 8 行时，有 2 个空位。当他们排列成 9 行时，有 3 个空位。如果乐队的成员数在 100 到 200 之间，那么乐队的成员总数是多少？

(A) 144 (B) 150 (C) 120 (D) 160 (E) 172

Problem 5

What day of the week will be 5000 days from Sunday?

问题 5

星期日的 5000 天后是星期几？

(A) Monday 星期一

(B) Tuesday 星期二

(C) Wednesday 星期三

(D) Thursday 星期四

(E) Friday 星期五

Problem 6

Determine how many two-digit numbers satisfy the following property: when the number is added to the number obtained by reversing its digits, the sum is 143.

问题 6

确定有多少个两位数满足以下条件：当该数字与其反转后的数字相加时，和为143。

(A) 5　　(B) 6　　(C) 7　　(D) 8　　(E) 10

Problem 7

The year 2002 is a palindrome (a number that reads the same from left to right as it does from right to left). What is the sum of the digits of the next year after 2002 that is a palindrome?

问题 7

2002 年是一个回文数年（从左到右和从右到左读起来相同的数称为回文数）。2002 年之后的下一个回文数年的数位之和是多少？

(A) 6　　(B) 7　　(C) 8　　(D) 10　　(E) 12

Problem 8

The product of the two 99-digit numbers 202, 020, 202, ⋯, 020, 202 and 707, 070, 707,⋯, 070, 707 has thousands digit A and units digit B. What is the product of A and B?

问题 8

两个 99 位数字 202，020，202，⋯，020，202 和 707，070，707，⋯，070，707 的乘积的千位数字是 A，个位数字是 B。A 和 B 的积是多少？

(A) 4　　(B) 6　　(C) 8　　(D) 10　　(E) 12

Problem 9

When the three integers 618, 343, and 277 are divided by a positive integer, d, where $d > 1$, the remainders are the same. What is the smallest possible value of d?

问题 9

当三个整数 618，343 和 277 被一个大于 1 的正整数 d 除时，余数相同。d 的最小可能值是多少？

(A) 5　　(B) 11　　(C) 25　　(D) 55　　(E) 16

Problem 10

Wendy noticed when she stacked her quarters in piles of 5 she had 3 left over and when she stacked them in piles of 7 she had 5 left over. If she has less than ten dollars' worth of quarters, how many quarters does she have?

问题 10

温迪发现，当她把 25 美分的硬币按 5 个一堆叠放时还剩下 3 个；按 7 个一堆叠放时还剩下 5 个。如果她的硬币总值少于 10 美元，那么她有多少个 25 美分的硬币？

(A) 11 (B) 22 (C) 33 (D) 44 (E) 55

Problem 11

Today is a Saturday in March. What day of the week will it be one year from today since next year is not a leap year?

问题 11

今天是三月的一个星期六。已知明年不是闰年，一年后是星期几？

(A) Monday 星期一

(B) Tuesday 星期二

(C) Wednesday 星期三

(D) Saturday 星期六

(E) Sunday 星期日

Problem 12

When 51 is divided by the positive integer k, the remainder is 3. For how many different values of k is this true?

问题 12

当 51 被正整数 k 除时，余数是 3。对于多少个不同的 k 值，这个条件成立？

(A) 10 (B) 8 (C) 7 (D) 4 (E) 5

Problem 13

Both m and n are integers. $5m+3n$ is divisible by 11. What is the remainder when $9m+n$ is divided by 11?

问题 13

m 和 n 都是整数。$5m+3n$ 能被 11 整除。那么 $9m+n$ 除以 11 的余数是多少？

(A) 0 (B) 1 (C) 4 (D) 7 (E) 9

Problem 14

What is the least positive integer value of p such that $7p$ divided by 11 has a remainder of 1?

问题 14

求最小的正整数 p，使得 $7p$ 除以 11 的余数为 1？

(A) 10　(B) 9　(C) 8　(D) 7　(E) 6

Problem 15

Find two integers between 1 and 100 such that for each:

a) if you divide by 4, the remainder is 3;

b) if you divide by 3, the remainder is 1;

c) if you divide by 5, the remainder is 1.

问题 15

找出 1 ～ 100 之间的两个整数，使得每个整数满足以下条件：

a) 除以 4 余 3；

b) 除以 3 余 1；

c) 除以 5 余 1。

(A) (11, 71)　(B) (31, 91)　(C) (21, 51)　(D) (16, 31)　(E) (46, 61)

Problem 16

When two different numbers are divided by 11, remainders of 7 and 9, respectively, are left. What is the remainder when the sum of these two numbers is divided by 11?

问题 16

当两个不同的数分别被 11 除时，余数分别为 7 和 9。那么这两个数的和除以 11 的余数是多少？

(A) 10　(B) 8　(C) 5　(D) 4　(E) 6

Chapter 4　Sequences and Series

第 4 章　数列与求和

4.1 Basic Knowledge 基础内容

4.1.1 Terms 项

An element of a sequence is called a term of the sequence, written as $a_1, a_2, a_3, \cdots$.

a_1 is called the first term.

a_n is called the general term or n^{th} term.

The sum of the first n terms is expressed as S_n. For example, S_{12} means the sum of the first twelve terms.

数列中的每个元素称为该数列的项，表示为 a_1，a_2，a_3，$\cdots$。

a_1 叫作首项，a_n 叫作通项或者第 n 项。

数列前 n 项的和用 S_n 表示。例如，S_{12} 表示数列前 12 项的和。

4.2 Arithmetic Sequences 等差数列

4.2.1 Definition 定义

- If any two consecutive terms in a sequence $a_1, a_2, a_3, \cdots, a_n, \cdots$, have the same difference, the sequence is called an arithmetic sequence (or arithmetic progression).

- The same difference is called the common difference (d).

- A finite sequence such as 1, 2, 3, 4, 5, $\cdots$, in which each term after the first is obtained by adding the preceding term by a fixed number, is an example of an arithmetic sequence.

- If we insert a number c between two numbers a and b such that a, c, b are in arithmetic sequence, then c is called the arithmetic mean of a and b and $c=\frac{(a+b)}{2}$.

- 如果一个数列 $a_1, a_2, a_3,\cdots, a_n,\cdots$ 中的任意两个连续项都具有相同的差值，则该数列称为等差数列。
- 相同的差叫作公差（用 d 表示）。
- 有限数列，例如 1, 2, 3, 4, 5,⋯是一个等差数列的例子，其中从第二项开始每个项，都是通过在前一项上加一个固定的数字得到的。
- 如果我们在 a 和 b 两个数之间插入一个数 c，使得 a, c, b 形成等差数列，则 c 被称为 a 和 b 的算术平均数，且 $c=\frac{(a+b)}{2}$。

4.2.2 The Common Difference of Arithmetic Sequences 等差数列的公差 (d)

$$d=a_{n+1}-a_n$$

$$d=\frac{a_m-a_n}{m-n}$$

$$\frac{d}{2}=\frac{\frac{S_m}{m}-\frac{S_n}{n}}{m-n}$$

4.2.3 The n th Term of Arithmetic Sequences 等差数列的通项

- The nth term is expressed as $a_n=a_1+(n-1)d$.
- Other forms: $a_n=a_m+(n-m)d$.
- $a_n=S_n-S_{n-1}$ (n is a positive integer and $n>1$).
- If $m+n=p+q$ where m, n, p, and q are positive integers, then $a_m+a_n=a_p+a_q$.
- If $m+n=2q$, then $a_m+a_n=2a_q$.
- 等差数列的第 n 项可以表示为 $a_n=a_1+(n-1)d$。
- 等差数列的第 n 项还可以表示为 $a_n=a_m+(n-m)d$。
- $a_n=S_n-S_{n-1}$（n 是正整数 $n>1$）。
- 如果 $m+n=p+q$，当 m，n，p 和 q 是正整数时，那么 $a_m+a_n=a_p+a_q$。
- 如果 $m+n=2q$，那么 $a_m+a_n=2a_q$。

Example 1

In an arithmetic sequence $a_1+a_3+a_5=105, a_2+a_4+a_6=99$. Find a_{20}.

例题 1

在等差数列中，$a_1+a_3+a_5=105$，$a_2+a_4+a_6=99$，求 a_{20}。

(A) −1　　(B) 1　　(C) 3　　(D) 7　　(E) −2

Solution

From $a_1+a_3+a_5=105$, we get $3a_3=105$, $a_3=35$. Similarly from $a_2+a_4+a_6=99$, we get $a_4=33$. The common difference of this arithmetic sequence is therefore $d=-2$. Thus, $a_{20}=a_4+(20-4)\times d$ = (B)1.

解　析

由 $a_1+a_3+a_5=105$ 可得 $3a_3=105$，$a_3=35$。同理，由 $a_2+a_4+a_6=99$，可得 $a_4=33$。因此这个等差数列的公差 $d=-2$。所以 $a_{20}=a_4+(20-4)\times d=1$，故选 (B)。

4.2.4 The Sum of *n* Terms in the Arithmetic Sequences 等差数列前 *n* 项的和

$$S=\frac{(a_1+a_n)n}{2},$$

$$S=na_1+\frac{(n-1)d}{2}n.$$

Example 2

On her first day of work, Anna sold two widgets. On day two, she sold four widgets. On day three, she sold six widgets, and on each succeeding day, she sold two more widgets than she had sold on the previous day. How many widgets in total had Anna sold after working 18 days?

例题 2

在工作第一天，安娜卖出了两件小工具。第二天，她卖出了四件小工具。第三天，她卖出了六件小工具。后面的每一天，她都比前一天多卖出两件小工具。经过 18 天，安娜总共卖出了多少件小工具？

(A) 36　　(B) 46　　(C) 210　　(D) 306　　(E) 342

Solution

The sum S_n of the first n terms of an arithmetic sequence is given by $S_n=\frac{n}{2}(a_1+a_n)$.

For n=18, $S_{18}=\frac{(2+36)\times 18}{2}=18\times 19=342$.

Therefore, the total number of widgets sold after 18 days is (E) 342.

解　析

等差数列前 n 项的和 S_n 公式为 $S_n=\frac{n}{2}(a_1+a_n)$。

当 $n=18$ 时，$S_{18}=\frac{(2+36)\times 18}{2}=18\times 19=342$。

因此，经过 18 天，总共卖出的小工具数量是 342，故选 (E)。

Example 3

The sum of six consecutive positive integers is 2025. What is the largest of these six integers?

例题 3

六个连续的正整数之和是 2025。这六个整数中最大的一个是多少？

(A) 335　　(B) 337　　(C) 338　　(D) 340　　(E) 342

Solution

Let the first term be x. Our integers are $x, x+1, x+2, x+3, x+4, x+5$.

We have, $6x+15=2025\Rightarrow x=335\Rightarrow x+5=\text{(D)}340$.

解　析

设第一项为 x。则连续的六个整数分别是 x，$x+1$，$x+2$，$x+3$，$x+4$，$x+5$。

根据题意，$6x+15=2025\Rightarrow x=335\Rightarrow x+5=340$，故选 (D)。

Example 4

Along the route of a bicycle race, 7 water stations are evenly spaced between the start and finish lines, as shown in the figure below. There are also 3 repair stations evenly spaced between the start and finish lines. The last repair station is located 4 miles after the 5th water station. How long is the race in miles?

例题 4

在自行车比赛的路线中，起点和终点之间均匀分布着 7 个水站，如下图所示。起点和终点之间还均匀分布着 3 个维修站。最后一个维修站位于第五个水站之后 4 英里。比赛的总长度是多少英里？

(A) 7 (B) 16 (C) 21 (D) 32 (E) 36

Solution

Suppose that the race is d miles long. The water stations are located at $\frac{d}{8},\frac{2d}{8},\cdots,\frac{7d}{8}$ miles from the start, and the repair stations are located at $\frac{d}{4},\frac{2d}{4},\frac{3d}{4}$ miles from the start.

We are given that $\frac{5d}{8}+4=\frac{3d}{4}$ from which

$$\frac{5d}{8}+4=\frac{6d}{8}$$

$$\frac{d}{8}=4$$

$$d=\text{(D)}32$$

解　析

假设比赛赛道长度为 d 英里。水站位于起点 $\frac{d}{8}$，$\frac{2d}{8}$，…，$\frac{7d}{8}$ 英里处，维修站位于起点 $\frac{d}{4}$，$\frac{2d}{4}$，$\frac{3d}{4}$ 英里处。

由题可得 $\frac{5d}{8}+4=\frac{3d}{4}$，解方程得

$$\frac{5d}{8}+4=\frac{6d}{8}$$
$$\frac{d}{8}=4$$
$$d=32$$

故选 (D)。

Example 5

An arithmetic sequence is a sequence in which each term after the first is obtained by adding a constant to the previous term. For example, 3, 6, 9, 12, 15 is an arithmetic sequence with five terms, in which the first term is 3 and the constant added is 3. Each row and each column in this 5 × 5 array is an arithmetic sequence with five terms. What is the value of X?

例题 5

一个等差数列是从第二项开始每个项，都是通过将一个常数加到前一项上得到的。例如，3，6，9，12，15 是一个包含 5 项的等差数列，其中第一项是 3，增加的常数是 3。在这个 5 × 5 的数列中，每一行和每一列都是一个包含 5 项的等差数列。X 的值是多少？

1				13
		X		
45				101

(A) 26　　(B) 31　　(C) 37　　(D) 40　　(E) 44

Solution

The middle term of the first row is $\frac{1+13}{2}=7$, since the middle number is just the average in an arithmetic sequence. Similarly, the middle of the bottom row is $\frac{45+101}{2}=73$. Applying this again for the middle column, the answer is $\frac{7+73}{2}=\text{(D)}40$.

解 析

第一行的中间项是 $\frac{1+13}{2}=7$，因为在项数为奇数项的等差数列，中间项为所有项的平均数。同样，底行的中间项是 $\frac{45+101}{2}=73$。对于中间列再次应用这一点，答案是 $\frac{7+73}{2}=40$, 故选 (D)。

4.3 Geometric Sequences 等比数列

4.3.1 Definition 定义

- If any two consecutive terms in a sequence $a_1, a_2, a_3,\cdots, a_n,\cdots$, have the same ratio, the sequence is called a geometric sequence (or geometric progression).
- The same ratio is called the common ratio (q or r):

$$q=\frac{a_{n+1}}{a_n}.$$

- A finite sequence such as 2, 4, 8, 16, 32,$\cdots$, in which each term after the first is obtained by multiplying the preceding term by a fixed number, is an example of a geometric sequence.
- If we insert a number c between two numbers a and b such that a, c, b are in geometric sequence, then c is called the geometric mean of a and b and $c^2=ab$, or $c=\pm\sqrt{ab}$, where $ab>0$.
- 如果一个序列 a_1，a_2，a_3，$\cdots$，a_n，$\cdots$ 中的任何两个连续项具有相同的比率，则该数列称为等比数列。
- 这个相同的比率叫作公比（用 q 或 r 表示）：

$$q=\frac{a_{n+1}}{a_n}。$$

- 2，4，8，16，32，$\cdots$ 是等比数列的一个例子，其中从第二项开始每个项，都是通过将前一项乘以一个固定数得到的。
- 如果我们在 a 和 b 两个数之间插入一个数 c，使得 a，c，b 形成等比数列，那么 c 被称为 a 和 b 的几何平均数，满足 $c^2=ab$，或 $c=\pm\sqrt{ab}$，这里 $ab>0$。

4.3.2 The Common Ratio of Geometric Sequences 等比数列的公比 (q)

$$q = \frac{a_{n+1}}{a_n}$$

4.3.3 The n th Term of Geometric Sequences 等比数列的第 n 项

$a_n = a_1 \cdot q^{n-1} \qquad (a_1 q \neq 0)$,

$a_n = a_m \cdot q^{n-m} \qquad (a_1 q \neq 0)$,

$a_n^2 = a_{n-k} \cdot a_{n+k} \qquad (n \geqslant k)$.

If $m+n=p+q$, $a_m \cdot a_n = a_p \cdot a_q$.

If $m+n=2t$, $a_m \cdot a_n = a_t^2$.

Example 6

Hulk can double the distance it jumps in each succeeding jump. If its first jump is 4 meters, the second jump is 8 meters, the third jump is 16 meters, and so on. On which jump will it first jump more than 1 kilometer?

例题 6

浩克每次跳跃的距离都会翻倍。如果他的第一次跳跃是 4 米，第二次跳跃是 8 米，第三次跳跃是 16 米，以此类推，那么他第一次能跳超过 1 千米是在第几次跳跃?

(A) 7th　　第 7 次

(B) 8th　　第 8 次

(C) 9th　　第 9 次

(D) 10th　　第 10 次

(E) 11th　　第 11 次

Solution

This is a geometric sequence in which the common ratio is 2. To find the jump that would be over a 1000 meters, we note that $2^{10}=1024$. However, because the first term is $2^2=4$, and $10-2+1=9$, therefore, the answer is (C)9th.

解 析

这是一个公比为 2 的等比数列。已知当 $2^{10}=1024$ 时，他跳跃超过了 1000 米。但数列的第一项是 $2^2=4$，由 $10-2+1=9$，故选 (C)。

Example 7

$\{a_n\}$ is a positive geometric sequence. Find the common ratio q if $a_{2010}=8a_{2007}$.

例题 7

$\{a_n\}$是一个正的等比数列，若 $a_{2010}=8a_{2007}$，求公比 q。

(A) 2 (B) 3 (C) 4 (D) 8 (E) 16

Solution

Since $\{a_n\}$ is a geometric sequence, we can get $\frac{a_{2010}}{a_{2007}}=q^3=8$. Thus, $q=\sqrt[3]{8}=(A)2$.

解 析

根据等比数列性质，我们得到 $\frac{a_{2010}}{a_{2007}}=q^3=8$。所以 $q=\sqrt[3]{8}=2$，故选 (A)。

Example 8

If five numbers are inserted between 8 and 5832, the fifth term in the geometric series formed is?

例题 8

将五个数插入在 8 和 5832 之间使其形成等比数列，第五项是多少？

(A) 648 (B) 832 (C) 1168

(D) 1944 (E) none of these 以上都不是

Solution

Denote the terms in the geometric sequence by $a_1 = 8, a_2 = 8r, \cdots, a_7 = 8r^6 = 5882$. Thus, $r^6 = \frac{5882}{8} = 729$, and $r = 3$. Therefore $a_5 = 8r^4 = (A)648$.

解 析

根据题意，表示等比数列中的项为 $a_1 = 8$，$a_2 = 8r$，$\cdots$，$a_7 = 8r^6 = 5882$。所以 $r^6 = 729$，$r = 3$。因此 $a_5 = 8r^4 = 648$，故答案为 (A)。

Example 9

In a sequence of positive integers, each term after the second is the product of the previous two terms. The sixth term is 25000. What is the first term?

例题 9

在一个正整数数列中，第二项之后的每个项都是前两项的乘积。第六项是 25 000。问第一项是多少?

(A) 1 (B) 2 (C) 4 (D) 8 (E) 10

Solution

Consider the first term is a and the second term is b. Then, the next items 3 to 6 are ab, ab^2, a^2b^3 and a^3b^5. Notice that $25000 = 2^3 \times 5^5$, then we obtain $a = 2$ and $b = 5$. The answer is (B).

解 析

设第一项为 a，第二项为 b。则接下来的第 3 ～ 6 项分别为 ab，ab^2，a^2b^3 和 a^3b^5。已知 $25000 = 2^3 \times 5^5$，因此 $a = 2$，$b = 5$，故答案为 (B)。

Example 10

A certain calculator has only two keys [+ 1] and [×2]. When you press one of the keys, the calculator automatically displays the result. For instance, if the calculator originally displayed “3” and you pressed [+1], it would display “4” . If you then pressed [×2], it would display “8” . Starting with the display “1” , what is the fewest number of keystrokes you would need to reach “205” ?

例题 10

某款计算器只有两个键 [+1] 和 [×2]。当你按下其中一个键时，计算器会自动显示结果。例如，如果计算器最初显示“3”，你按下 [+1]，它会显示“4”。如果你接着按 [×2]，它会显示“8”。从显示“1”开始，达到显示“205”所需的最少按键次数是多少？

(A) 8 (B) 9 (C) 10 (D) 11 (E) 12

Solution

We can start at 205 and work our way down to 1. We want to press the button that “multiplies by 2” the most, but since we are going down instead of up, we “divide by 2” instead. If we come across an odd number, then we will subtract that number by 1. If we have an even number, then we divide the number by 2. We get:

205, 204, 102, 51, 50, 25, 24, 12, 6, 3, 2, 1.

It’s clear that the answer should be (D)11.

解 析

反向计算，我们从 205 到 1。因为是反向，按键“乘以 2”会变成“除以 2”。若遇奇数，则减去 1。若遇偶数，则除以 2。根据以上运算规则可得

205，204，102，51，50，25，24，12，6，3，2，1。

显然，答案应是 (D)。

PRACTICE 练习

Problem 1

In an arithmetic sequence a_n, if $a_1 + a_9 = 10$, find a_5.

问题 1

在一个等差数列 a_n 中，如果 $a_1 + a_9 = 10$，求 a_5。

(A) 5 (B) 6 (C) 8 (D) 10 (E) 20

Problem 2

The second and fourth terms of a geometric sequence are 2 and 6. Which of the following is a possible first term?

问题 2

一个等比数列的第二项和第四项分别为 2 和 6。以下哪个是可能的第一项?

(A) $-\sqrt{3}$ (B) $-\frac{2\sqrt{3}}{3}$ (C) $-\frac{2\sqrt{3}}{3}$ (D) $\sqrt{3}$ (E) 3

Problem 3

What is the value of

$2+4+6+\cdots+2016+2018-1-3-5-\cdots-2017-2019$?

问题 3

表达式$2+4+6+\cdots+2016+2018-1-3-5-\cdots-2017-2019$ 的值为多少?

(A) -1010 (B) -1009 (C) -1004 (D) 1004 (E) 1009

Problem 4

What is the value of the expression $\frac{1\times2\times3\times4\times5\times6\times7\times8\times9}{1+2+3+4+5+6+7+8+9}$?

问题 4

表达式 $\frac{1\times2\times3\times4\times5\times6\times7\times8\times9}{1+2+3+4+5+6+7+8+9}$ 的值为多少?

(A) 2040 (B) 3460 (C) 5268 (D) 6486 (E) 8064

Problem 5

A sequence of numbers starts with 2, 3, and 4. The fourth number of the sequence is the sum of the previous three numbers in the sequence: $2+3+4=9$. In the same way, every number after the fourth is the sum of the previous three numbers. What is the seventh number in the sequence?

问题 5

一个数字数列以 2，3 和 4 开始。数列的第四个数字是前面三个数字的和：$2+3+4=9$。以同样的方式，第四个数字之后的每个数字都是前面三个数字的和。那么数列中的第七个数字是什么？

(A) 8　(B) 17　(C) 36　(D) 54　(E) 68

Problem 6

To complete the grid below, each of the digits 1 through 4 must occur once in each row and once in each column. What number will occupy the lower right-hand square?

问题 6

为了完成下面的网格，数字 1 ～ 4 必须在每一行和每一列中各出现一次。那么右下角的方格中应填哪个数？

4	3		
	2		
3			
		1	

(A) 1　(B) 2　(C) 3

(D) 4　(E) cannot be determined 无法确定

Problem 7

The numbers −4, 1, 2, 3 and 7 are rearranged according to these rules:

(1) The largest isn’t first, but it is in one of the first three places.

(2) The smallest isn’t last, but it is in one of the last three places.

(3) The median isn’t first or last.

What is the average of the first and last numbers?

问题 7

数字 −4，1，2，3 和 7 按照以下规则重新排列：

(1) 最大的数字不是第一个，但在前三个位置之一。

(2) 最小的数字不是最后一个，但在最后三个位置之一。

(3) 中位数既不是第一个也不是最后一个。

那么，第一个和最后一个数字的平均值是多少？

(A) −1.5　(B) 1.5　(C) 2　(D) 4.5　(E) 5

Problem 8

Mr. Peter needs to know the combined weight in pounds of three boxes he wants to mail. However, the only available scale is not accurate for weights less than 200 pounds or more than 250 pounds. So the boxes are weighed in pairs in every possible way (weighing two boxes at a time). The results are 218, 237 and 249 pounds. What is the combined weight in pounds of the three boxes?

问题 8

彼得先生需要知道他想邮寄的三个箱子的总重量（以磅为单位）。然而，唯一可用的秤对于低于 200 磅或超过 250 磅的重量不准确。因此，这些箱子以所有可能的方式成对称重（每次称两个箱子）。结果分别是 218 磅，237 磅和 249 磅。那么这三个箱子的总重量是多少磅？

(A) 308　　(B) 329　　(C)337　　(D) 352　　(E) 367

Problem 9

Al the monkey ate 105 delicious yellow bananas from June 1st through June 6th. Each day he ate three more bananas than those on the previous day. How many delicious bananas did Al eat on June 5th?

问题 9

猴子阿尔从 6 月 1 日—6 月 6 日吃了 105 根美味的黄香蕉。每天比前一天多吃 3 根香蕉。那么阿尔在 6 月 5 日吃了多少根美味的香蕉？

(A) 15　　(B) 17　　(C) 18　　(D) 22　　(E) 25

Problem 10

In an arithmetic sequence $\{a_n\}$, if $a_2+a_{10}=20$, find a_6.

问题 10

在一个等差数列 $\{a_n\}$ 中，如果 $a_2+a_{10}=20$，求 a_6。

(A) 5　　(B) 6　　(C) 8　　(D) 10　　(E) 20

Problem 11

Find the sum of all counting numbers between 50 and 350 that have 1 as the last digit.

问题 11

求所有在 50 ～ 350 之间、以 1 为最后一位数字的自然数的总和。

(A) 5050　　(B) 5880　　(C) 5441　　(D) 5552　　(E) 5338

Problem 12

By adding the same constant to each of 20, 50, 100 a geometric progression results. The common ratio is?

问题 12

通过对 20，50 和 100 加上相同的常数，可以得到一个等比数列。问公比是多少？

(A) $\frac{5}{3}$　(B) $\frac{4}{3}$　(C) $\frac{3}{2}$　(D) $\frac{1}{2}$　(E) $\frac{1}{3}$

Problem 13

Find the value of the expression: $99-97+95-93+91-89+\cdots+7-5+3$.

问题 13

求表达式 $99-97+95-93+91-89+\cdots+7-5+3$ 的值。

(A) 40　(B) 43　(C) 50　(D) 51　(E) 53

Problem 14

What is the product of $\frac{2}{3}\times\frac{3}{4}\times\frac{4}{5}\times\cdots\times\frac{2023}{2024}$?

问题 14

求 $\frac{2}{3}\times\frac{3}{4}\times\frac{4}{5}\times\cdots\times\frac{2023}{2024}$ 的积。

(A) 2023　(B) 1　(C) $\frac{1}{2024}$　(D) $\frac{1}{2023}$　(E) $\frac{1}{1012}$

Problem 15

What is the value of $5\times(1-2+3-4+5-6+\cdots+999-1000)$?

问题 15

求 $5\times(1-2+3-4+5-6+\cdots+999-1000)$ 的值。

(A) -2500　(B) -2000　(C) -500　(D) 500　(E) 2000

Chapter 5 Ratios, Rates, and Proportions

第5章 比率、速率和比例

5.1 Basic Knowledge 基础内容

5.1.1 Ratios 比率

Ratios are used to compare two or more numbers.

For any two numbers a and b ($b \neq 0$), the ratio is written as

$$a:b = a \div b = \frac{a}{b} = a/b$$

比率用于比较两个或多个数。

对于任何两个数 a 和 b（$b \neq 0$），比率可以表示为

$$a:b = a \div b = \frac{a}{b} = a/b$$

5.1.2 Rates 速率

A rate is a ratio used to compare two numbers of different units. If the second term of the ratio is 1, the rate is called a unit rate.

速率是用于比较不同单位的两个数字的比率。如果比率的第二项为 1，则称为单位速率。

5.1.3 Proportions 比例

A proportion is a part, share or number considered in comparative relation to a whole. For example, there are several pieces of fruit, and 1 in every 5 of those is an apple. Then the proportion of apple is $\frac{1}{5}$ or 20%.

比例形容某一部分和整体之间的数量关系。例如，有一些水果，每 5 个中就有 1 个苹果，那么苹果的比例为 $\frac{1}{5}$ 或 20%。

Example 1

A lake contains 180 trout, along with a variety of other fish. When a marine biologist catches and releases a sample of 220 fish from the lake, 20 are identified as trout. Assume that the proportion of number of trout to that of all fish is the same in both the sample and the lake. What is the number of fish in the lake?

例题 1

湖里有 180 条鳟鱼和各种其他鱼类。一名海洋生物学家捕获并释放了 220 条鱼样本，其中 20 条被确认为鳟鱼。假设鳟鱼与总鱼数的比率在样本和湖中相同。问湖中共有多少条鱼?

(A) 1560 (B) 1680 (C) 1750 (D) 1800 (E) 1980

Solution

Note that $\frac{\text{number of trout}}{\text{total number of fish}}=\frac{20}{220}=\frac{1}{11}$. So, the total number of fish is 11 times that of trout. Since the lake contains 180 trout, there are $11\times180=1980$ fish in the lake. The answer is (E).

解析

样本中 $\frac{\text{鳟鱼数}}{\text{鱼的总数}}=\frac{20}{220}=\frac{1}{11}$。所以鱼的总数是鳟鱼数的 11 倍。湖中有 180 条鳟鱼，湖中鱼的总数应为 $11\times180=1980$ 条。故答案是 (E)。

Example 2

Harry made a strawberry pie for a picnic. He ate only $\frac{1}{3}$ of the pie, and left the rest to his friends. A hedgehog came by and ate $\frac{1}{4}$ of what Harry left behind. After that, a chick ate $\frac{1}{5}$ of what the hedgehog left behind. How much of the original pie remained after the chick left?

例题 2

哈利做了一个草莓派去野餐。他只吃了 $\frac{1}{3}$ 的派，剩下的留给朋友们。一只刺猬经过并吃掉了哈利剩下的 $\frac{1}{4}$。之后，一只小鸡吃掉了刺猬剩下的 $\frac{1}{5}$。问小鸡吃完后，草莓派还剩下多少？

(A) $\frac{1}{5}$　(B) $\frac{1}{4}$　(C) $\frac{1}{3}$　(D) $\frac{2}{5}$　(E) $\frac{7}{12}$

Solution

Note that Harry ate $\frac{1}{3}$ of the pie. After that, $1-\frac{1}{3}=\frac{2}{3}$ of the pie was left behind.

The hedgehog ate $\frac{1}{4}\times\frac{2}{3}=\frac{1}{6}$ of the pie. After that, $\frac{2}{3}-\frac{1}{6}=\frac{1}{2}$ of the pie was left behind.

The chick ate $\frac{1}{5}\times\frac{1}{2}=\frac{1}{10}$ of the pie. After that, $\frac{1}{2}-\frac{1}{10}=\frac{2}{5}$ of the pie was left behind.

More simply, we can condense the solution above into the following equation:

$$\left(1-\frac{1}{3}\right)\left(1-\frac{1}{4}\right)\left(1-\frac{1}{5}\right)=\frac{2}{5}$$

Therefore, the answer is (D).

解　析

请注意：哈利吃了 $\frac{1}{3}$。之后，剩下了 $1-\frac{1}{3}=\frac{2}{3}$。

刺猬吃了 $\frac{1}{4}\times\frac{2}{3}=\frac{1}{6}$。之后，剩下了 $\frac{2}{3}-\frac{1}{6}=\frac{1}{2}$。

小鸡吃了 $\frac{1}{5}\times\frac{1}{2}=\frac{1}{10}$。之后，剩下了 $\frac{1}{2}-\frac{1}{10}=\frac{2}{5}$。

可将上述过程简化为以下算式：$\left(1-\frac{1}{3}\right)\left(1-\frac{1}{4}\right)\left(1-\frac{1}{5}\right)=\frac{2}{5}$。故选 (D)。

5.2 Important Properties 重要性质

5.2.1 Properties of Ratios 比率的性质

Property 1 The first term of a ratio can be any number. The second term can also be any number except zero.

性质 1 比率的第一项可以是任意数。第二项可以是任意非零数。

Property 2 If the two terms are multiplied by the same number d, the ratio does not change.

$$a:b=(a\times d):(b\times d)\Rightarrow 3:7=(3\times 5):(7\times 5)=15:35$$

性质 2 如果比率的两项同时乘以相同的数 d，比率不变。

$$a:b=(a\times d):(b\times d)\Rightarrow 3:7=(3\times 5):(7\times 5)=15:35$$

Property 3 If the two terms are divided by the same number c ($c\neq 0$), the ratio does not change.

$$a:b=(a\div c):(b\div c)\Rightarrow 10:15=(10\div 5):(15\div 5)=2:3$$

性质 3 如果比率的两项同时除以相同的数 c ($c\neq 0$)，比率不变。

$$a:b=(a\div c):(b\div c)\Rightarrow 10:15=(10\div 5):(15\div 5)=2:3$$

Property 4 If the total number of parts is $m=A+B$, and $A:B=a:b$, then the fractional part of a is $\dfrac{a}{a+b}$, and the fractional part of b is $\dfrac{b}{a+b}$.

$$A=\frac{a}{a+b}\times m \text{, and } B=\frac{b}{a+b}\times m$$

性质 4 如果总数为 $m=A+B$，且 $A:B=a:b$，那么 a 的分数部分是 $\dfrac{a}{a+b}$，b 的分数部分是 $\dfrac{b}{a+b}$。

$$A=\frac{a}{a+b}\times m\text{，且 } B=\frac{b}{a+b}\times m$$

Property 5 If the total number of parts is $m=A+B+C$, and $A:B:C=a:b:c$, then the fractional part of a is $\dfrac{a}{a+b+c}$, the fractional part of b is $\dfrac{b}{a+b+c}$, and the fractional part of c is $\dfrac{c}{a+b+c}$.

$$A=\frac{a}{a+b+c}\times m\text{, } B=\frac{b}{a+b+c}\times m\text{, and } C=\frac{c}{a+b+c}\times m$$

性质 5 如果部分总数为 $m=A+B+C$，且 $A:B:C=a:b:c$，那么 a 的分数部分是 $\frac{a}{a+b+c}$，b 的分数部分是 $\frac{b}{a+b+c}$，c 的分数部分是 $\frac{c}{a+b+c}$。

$$A=\frac{a}{a+b+c}\times m,\ B=\frac{b}{a+b+c}\times m,\ C=\frac{c}{a+b+c}\times m$$

Example 3

Suppose 35% of x equals 25% of y. What percentage of x is y?

例题 3

假设 x 的 35% 等于 y 的 25%，那么 y 是 x 的百分之多少？

(A) 25 (B) 80 (C) 105 (D) 120 (E) 140

Solution

According to the condition, we have $25\%y = 35\%x$. Since $25\%=\frac{1}{4}$, multiplying both sides of the equation by 4 shows that y is $35\times4=(\text{E})140$ percent of x.

解 析

根据题目所给等量关系，可列方程 $25\%y = 35\%x$。因 $25\%=\frac{1}{4}$，所以方程两边同乘以 4 得 y 是 x 的 $35\%\times4=140\%$，故选 (E)。

Example 4

A group of frogs is living in a tree. A frog turns green when in the shade and turns yellow when in the sun. Initially, the ratio of green to yellow frogs was 5 : 2. Then 4 green frogs moved to the sunny side and 7 yellow frogs moved to the shady side. Now the ratio is 3 : 1. What is the difference between the number of green frogs and the number of yellow frogs now?

例题 4

一群青蛙生活在树上。青蛙在阴凉处变成绿色，在阳光下变成黄色。最初，绿色青蛙与黄色青蛙的比率是 5 : 2。然后，4 只绿色青蛙移动到阳光下，7 只黄色青蛙移动到阴凉处。现在，绿色与黄色青蛙的比率是 3 : 1。问现在绿色青蛙与黄色青蛙的数量差是多少？

(A) 34 (B) 36 (C) 42 (D) 48 (E) 66

Solution

Let the initial numbers of green frogs and yellow frogs be g and y, respectively. Since the initial ratio of the number of green frogs to that of yellow frogs is 5 : 2, we have $2g=5y$. Then, 4 green frogs moved to the sunny side and 7 yellow frogs moved to the shady side, the current numbers of green frogs and yellow frogs are $g+3$ and $y-3$, respectively. We are given that $\frac{g+3}{y-3}=\frac{3}{1}$, so $g+3=3y-9$. Since $2g=5y$, we can get $5y+6=6y-18$, so $y=24$ and $g=60$. Therefore, there are $60+3$=63 green frogs and $24-3=21$ yellow frogs now, and the difference is $63-21=\text{(C)}42$.

解 析

设最初绿色青蛙的数量为 g，黄色青蛙的数量为 y。因为绿色青蛙与黄色青蛙的初始比率是 5 : 2，所以 $2g=5y$。现在，4 只绿色青蛙移动到了阳光一侧 7 只黄色青蛙移动到了阴影一侧，因此绿色青蛙的新数量为 $g+3$，黄色青蛙的新数量为 $y-3$，于是可以得到新的比率是 3 : 1。因此 $\frac{g+3}{y-3}=\frac{3}{1}$，所以 $g+3=3y-9$。又因为 $2g=5y$，代入方程得到 $5y+6=6y-18$，解得 $y=24$，$g=60$。所以，绿色青蛙现在的数量为 $60+3$=63 只，黄色青蛙的数量为 $24-3=21$ 只，差值为 $63-21=42$ 只，故选 (C)。

5.2.2 Properties of Proportion 比例的性质

Property 6 $\frac{a}{b}=\frac{c}{d}$ is equivalent to $ad=bc$, $\frac{a}{c}=\frac{b}{d}$, $\frac{d}{b}=\frac{c}{a}$, $\frac{b}{a}=\frac{d}{c}$.

性质 6 $\frac{a}{b}=\frac{c}{d}$ 等价于 $ad=bc$，$\frac{a}{c}=\frac{b}{d}$，$\frac{d}{b}=\frac{c}{a}$，$\frac{b}{a}=\frac{d}{c}$。

$\frac{8}{6}=\frac{4}{3}\Rightarrow 8\times3=4\times6\Rightarrow\frac{8}{4}=\frac{6}{3}\Rightarrow\frac{3}{6}=\frac{4}{8}\Rightarrow\frac{6}{8}=\frac{3}{4}$。

Property 7 If $\frac{a}{b}=\frac{c}{d}$, then $\frac{a+b}{b}=\frac{c+d}{d}$ and $\frac{a-b}{b}=\frac{c-d}{d}$.

性质 7 如果 $\dfrac{a}{b}=\dfrac{c}{d}$，那么 $\dfrac{a+b}{b}=\dfrac{c+d}{d}$，$\dfrac{a-b}{b}=\dfrac{c-d}{d}$。

$\dfrac{21}{7}=\dfrac{30}{10}\Rightarrow\dfrac{21+7}{7}=\dfrac{30+10}{10}(=4)$，且 $\dfrac{21-7}{7}=\dfrac{30-10}{10}(=2)$。

Property 8 If $\dfrac{a}{b}=\dfrac{c}{d}$, then $\dfrac{a+b}{a-b}=\dfrac{c+d}{c-d}$.

性质 8 如果 $\dfrac{a}{b}=\dfrac{c}{d}$，那么 $\dfrac{a+b}{a-b}=\dfrac{c+d}{c-d}$。

$\dfrac{21}{7}=\dfrac{30}{10}\Rightarrow\dfrac{21+7}{21-7}=\dfrac{30+10}{30-10}\left(\dfrac{28}{14}=\dfrac{40}{20}=2\right)$。

Property 9 If $\dfrac{a_1}{b_1}=\dfrac{a_2}{b_2}=\dfrac{a_3}{b_3}=\cdots=\dfrac{a_n}{b_n}$, then

$$\frac{a_1}{b_1}=\frac{a_1+a_2}{b_1+b_2}=\frac{a_1+a_2+a_3}{b_1+b_2+b_3}=\frac{a_1+a_2+a_3+\cdots+a_n}{b_1+b_2+b_3+\cdots+b_n}$$

性质 9 如果 $\dfrac{a_1}{b_1}=\dfrac{a_2}{b_2}=\dfrac{a_3}{b_3}=\cdots=\dfrac{a_n}{b_n}$，那么

$$\frac{a_1}{b_1}=\frac{a_1+a_2}{b_1+b_2}=\frac{a_1+a_2+a_3}{b_1+b_2+b_3}=\frac{a_1+a_2+a_3+\cdots+a_n}{b_1+b_2+b_3+\cdots+b_n}$$

Example 5

A shop advertises that everything has a 40% discount in today's sale. In addition, a coupon gives a 30% discount on sale prices. Using the coupon, what is the total percentage discount off the original price?

例题 5

一家商店打广告称“今日全场六折”。此外，还有一张优惠券可再打 7 折。那么使用优惠券后，价格比原价便宜了百分之几？

(A) 12 (B) 30 (C) 42 (D) 58 (E) 70

Solution

Let the original price of an item be x.

First, everything is 40% off, so the price is now $0.6x$.

Next, the extra coupon applies 30% off on the sale price, so the price after this discount will be $100\% - 30\% = 70\%$ of what it is now.

(Notice that this is not applied to the original price.)

$$70\% \times 0.6x = 0.42x$$

So, the price of the item after all discounts applied is $0.42x = 42\% \cdot x$. However, we need to find the percentage off the original price, not the current percentage of the original price. We then subtract 42 from 100 to find the answer (D)58.

解析

设物品的原价为 x。

首先，全部商品打六折，因此现在的价格为 $0.6x$。

接下来，优惠券额外提供 30% 的折扣，这个折扣是基于折后价，而不是原价，因此新的价格是之前价格的 $100\% - 30\% = 70\%$。

（注意，这个折扣不是应用在原价上的，不可直接将折扣按照 40% + 30% = 70% 来计算。）

$$70\% \times 0.6x = 0.42x$$

因此最终折扣后的价格是 $0.42x = 42\% \cdot x$，即只需要原价的 42% 就能购买商品，相当于得到 58% 的折扣，故选 (D)。

5.2.3 Continued Ratio 连比

The ratio of three or more quantities is called the continued ratio. For example, $a : b : c$ is a combinations of three separated ratios $\Rightarrow$ $a : b$, $a : c$, and $b : c$.

三个或三个以上的量的比率叫作连续比率，简称连比。例如，$a : b : c$ 表示三个比率的组合：$a : b$，$a : c$，$b : c$。

Property 10 If $a:b:c = m:n:t$, then $a:b = m:n$, $b:c = n:t$ and $c:a = t:m$.

For example, if $a:b:c = 2:3:4$, then $a:b = 2:3$, $b:c = 3:4$, and $c:a = 4:2$.

性质 10 如果 $a:b:c = m:n:t$，那么 $a:b = m:n$，$b:c = n:t$，$c:a = t:m$。

例如，如果 $a:b:c = 2:3:4$，那么 $a:b = 2:3$，$b:c = 3:4$，$c:a = 4:2$。

Property 11 If $a:b = m:n$, $b:c = n:t$, and $c:a = t:m$, then $a:b:c = m:n:t$.

For example, if $a:b = 2:3$, $b:c = 3:4$, and $c:a = 4:2$, then $a:b:c = 2:3:4$.

性质 11 如果 $a:b = m:n$，$b:c = n:t$，$c:a = t:m$，那么 $a:b:c = m:n:t$。

例如，如果 $a:b = 2:3$，$b:c = 3:4$，$c:a = 4:2$，那么 $a:b:c = 2:3:4$。

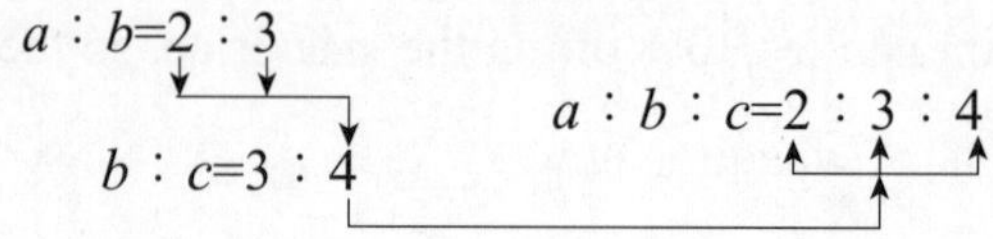

Property 12 If $a:b=m:n$, and $b:c=s:t$ (note $n \neq s$), then

$$a:b:c=(m\times s):(n\times s):(n\times t)$$

For example, if $a:b=2:3$, and $b:c=5:4$ (note $3 \neq 5$), then

$$a:b:c=(2\times 5):(3\times 5):(3\times 4)=10:15:12$$

性质 12 如果 $a:b=m:n$， $b:c=s:t$（注意 $n\neq s$），那么

$$a:b:c=(m\times s):(n\times s):(n\times t)$$

例如，如果 $a:b=2:3$， $b:c=5:4$（注意 $3 \neq 5$），那么

$$a:b:c=(2\times 5):(3\times 5):(3\times 4)=10:15:12$$

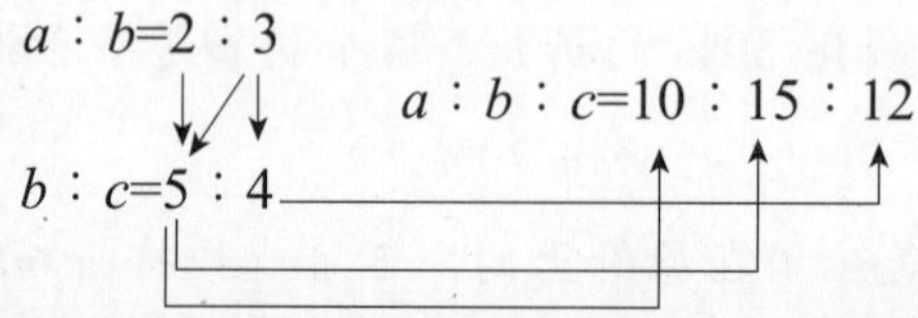

Example 6

If the degree measures of the angles of a triangle are in the ratio 2 : 2 : 5, what is the degree measure of the largest angle of the triangle?

例题 6

如果一个三角形的角度比率是 2 : 2 : 5，问该三角形的最大角度是多少？

(A) 24 (B) 36 (C) 54 (D) 81 (E) 100

Solution

Since the sum of the angles of a triangle is $180°$ and the ratio of the three angles is 2 : 2 : 5, we know the $180°$ is evenly divided into $2+2+5=9$ parts, and two angles have 2 parts respectively, while the other has 5 parts. Since 1 part is measured as $180° \div 9 = 20°$, the maximum angle is $20° \times 5 = 100°$. The answer is (E).

解 析

由于三角形的内角和为 $180°$，而三个角的角度比率为 2 : 2 : 5，即 $180°$ 总共分成了 2+2+5=9 份，有两个角各占了 2 份，第三个角占了 5 份。因为一份是 $180° \div 9 = 20°$，问题要求找出最大的角度，即 $20° \times 5 = 100°$。因此答案是 (E)。

5.3 Speed 速度

5.3.1 Basic Formulas Related to Distance, Rate, and Time 与距离、速率和时间相关的基本公式

$$d = rt$$

Where d stands for distance (given in units such as miles, feet, kilometers, meters, etc.), t stands for time (hours, minutes, seconds, etc.) and r stands for the (uniform, fixed, steady, constant, or average) rates of speed.

$$r = \frac{d}{t}$$

$$t = \frac{d}{r}$$

In the following figure, we add d_1 and d_2 to get the total distance $d: d = d_1 + d_2$, and we add t_1 and t_2 to get the total time $t: t = t_1 + t_2$.

However, we do not add r_1 and $r_2: r \neq r_1 + r_2$.

d 代表距离（以英里、英尺、公里、米等为单位），t 代表时间（小时、分钟、秒等），而 r 代表速度的速率（可以是均匀的、固定的、稳定的、恒定的或平均的）。

$$r = \frac{d}{t}$$

$$t = \frac{d}{r}$$

在下图中，我们将 d_1 和 d_2 相加得到总距离：$d: d = d_1 + d_2$，将 t_1 和 t_2 相加得到总时间：$t: t = t_1 + t_2$。

然而，我们不能将 r_1 与 r_2 相加：$r \neq r_1 + r_2$。

$$r = \frac{d}{t} = \frac{d_1 + d_2}{t_1 + t_2} = \frac{d_1 + d_2}{\dfrac{d_1}{r_1} + \dfrac{d_2}{r_2}} = \frac{r_1 r_2 (d_1 + d_2)}{d_1 r_2 + d_2 r_1}$$

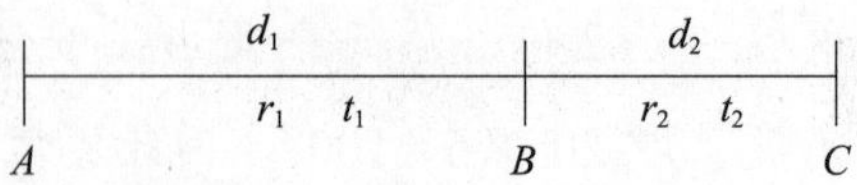

A–B	r_1	×	t_1	=	d_1
			+		+
B–C	r_2	×	t_2	=	d_2
			=		=
A–C	r	×	t	=	d

$$d_1 = d_2 \Rightarrow r = \frac{r_1 r_2 (d_1 + d_2)}{d_1 r_2 + d_2 r_1} = \frac{2 d r_1 r_2}{d(r_2 + r_1)} = \frac{2 r_1 r_2}{r_1 + r_2}$$

Example 7

NASA's Rocket was launched on July 30, 2020. After traveling 401248937 miles, it landed on Mars in Jezero Crater about 8.5 months later. Which of the following is closest to the Rover's average interplanetary speed in miles per hour?

例题 7

美国国家航空航天局的火箭载着火星探测器于 2020 年 7 月 30 日发射，飞行 401248937 英里后，约 8.5 个月后在火星上杰泽罗陨石坑着陆。以下哪个选项最接近探测器的平均星际速度？（单位：英里 / 小时）

(A) 7000 (B) 14000 (C) 70000 (D) 120000 (E) 670000

Solution

Note that 8.5 months is approximately $8.5 \times 30 \times 24$ hours. Therefore, the speed (in miles per hour) is:

$$\frac{401248937}{8.5 \times 30 \times 24} \approx \frac{400000000}{8.5 \times 30 \times 24} = \frac{10000000}{8.5 \times 3 \times 6} = \frac{10000000}{25.5 \times 6} \approx \frac{10000000}{150} \approx \text{(C)}70000$$

解 析

8.5 个月大约为 $8.5 \times 30 \times 24$ 小时。因此，速度（单位：英里 / 小时）为

$$\frac{401248937}{8.5 \times 30 \times 24} \approx \frac{400000000}{8.5 \times 30 \times 24} = \frac{10000000}{8.5 \times 3 \times 6} = \frac{10000000}{25.5 \times 6} \approx \frac{10000000}{150} \approx 70000$$

故选 (C)。

Example 8

Victor walks a mile to get to school each day. His route consists of 10 city blocks of equal length and he takes 4 minutes to walk each block. Today, after walking 5 blocks, Victor needs to make a detour, walking 3 blocks of equal length instead of 1 block to reach the next corner. From the time he starts his detour, at what speed, in mph, must he walk, in order to get to school at his usual time?

例题 8

维克多每天步行一英里去学校。他的路线由 10 个等长的街区组成，他每穿过一个街区需要 4 分钟。今天，走完 5 个街区后，维克多发现他必须绕行，改走 3 个等长的街区，而不是原来的 1 个街区，才能到达下一个拐角。从他开始绕行的那一刻起，他必须以每小时多少英里的速度行走，才能按平时的时间到达学校？

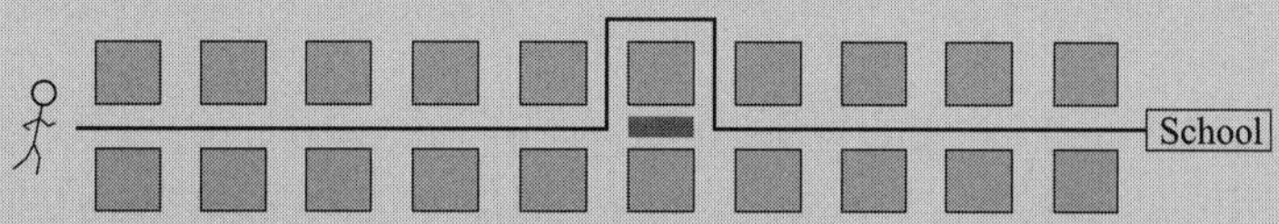

(A) 1.8　(B) 2.1　(C) 3　(D) 3.2　(E) 3.6

Solution

From the time Victor takes his detour, he must travel 7 blocks instead of 5, so the distance he travels is $0.1\times 7=0.7$ miles. Also, Victor must arrive at the school in $\frac{4\times 10}{2\times 60}=\frac{1}{3}$ hours. Thus, our final equation is $0.7\div\frac{1}{3}=0.7\times 3=$ (B)2.1.

解 析

根据题意可知，从维克多开始绕行的那一刻起，他必须走 7 个街区，而不是 5 个街区，所以路程为 $0.1\times 7=0.7$ 英里。另外，维克多必须在 $\frac{4\times 10}{2\times 60}=\frac{1}{3}$ 小时内到达学校，所以最终的算式是 $0.7\div\frac{1}{3}=0.7\times 3=2.1$，故选 (B)。

Example 9

After school, Mary and Nancy headed to the beach 6 miles away. Mary decided to walk while Nancy rode a bike. The graph below shows their journeys, indicating the time and distance traveled. What was the difference, in miles per hour, of Nancy's and Mary's average speeds?

例题 9

放学后，玛丽和南希前往 6 英里外的海滩。玛丽决定走路而南希骑自行车过去。下图显示了她们的行程，标明了所用时间和经过的路程。问南希和玛丽的平均速度相差多少英里每小时?

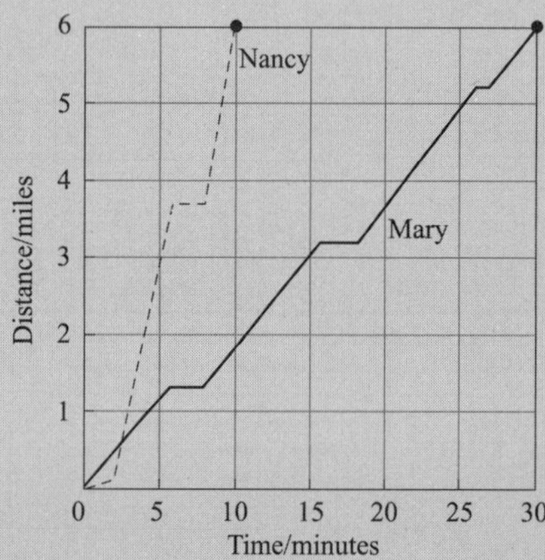

(A) 16　　(B) 18　　(C) 20　　(D) 24　　(E) 26

Solution

Nancy traveled 6 miles in 10 minutes. Since 10 minutes equals to $\frac{1}{6}$ hours and the average speed = $\frac{\text{distance}}{\text{time}}$, her average speed was $6 \div \frac{1}{6} = 36$ mph. Similarly, Mary spent 30 minutes walking 6 miles, so her average speed was 12 mph. Therefore, the answer is $36 - 12 = \text{(D)}24$.

解　析

南希在 10 分钟内骑行了 6 英里。10 分钟相当于 $\frac{1}{6}$ 小时，根据速度 = $\frac{\text{路程}}{\text{时间}}$，她

的速度是 $6 \div \frac{1}{6} = 36$ 英里每小时。同理，玛丽用 30 分钟走了 6 英里，所以她的平均速度是 12 英里每小时。因此，答案是 $36 - 12 = 24$，故选 (D)。

Example 10

Each of four days, Linda traveled for 180 minutes at a speed with which she spent an integer number of minutes for one mile. Each day after the first, her speed decreased so that the number of minutes to travel one mile increased by 15 minutes than the preceding day. In each day, her traveled distance was also an integer. What was the total number of miles for the four trips?

例题 10

在 4 天里，琳达每天花 180 分钟旅行。她每走 1 英里所需的时间为整数分钟。每天下来，她的速度都会下降，因此后一天每英里所需的时间比前一天增加 15 分钟。在 4 天中，她的旅行距离也都是整数英里。问 4 天旅行的总里程是多少？

(A) 15　　(B) 25　　(C) 40　　(D) 50　　(E) 65

Solution

Assume Linda's speed in the first day to be 1 mile per x minutes. Then the distance she traveled was $\frac{180}{x}$ miles in the first day. The distance for the following days are: $\frac{180}{x+15}, \frac{180}{x+30}, \frac{180}{x+45}$. Since the distance of each day is an integer, we know that $x, x+15, x+30, x+45$ are factors of 180. Therefore, $x = 15$ is the only solution, and

$$\frac{180}{15} + \frac{180}{30} + \frac{180}{45} + \frac{180}{60} = 12 + 6 + 4 + 3 = \text{(B)}25$$

解　析

假设琳达第一天的速度是每 x 分钟走 1 英里，可得第一天的路程为 $\frac{180}{x}$。同理，接下来几天的路程分别是 $\frac{180}{x+15}$，$\frac{180}{x+30}$，$\frac{180}{x+45}$。由于每天的路程都是整数，且 x，$x+15$，$x+30$，$x+45$ 都是 180 的因数，可得 $x = 15$ 为唯一解。所以总路程为

$$\frac{180}{15}+\frac{180}{30}+\frac{180}{45}+\frac{180}{60}=12+6+4+3=25$$

故选 (B)。

5.3.2 Two people walking in the same direction 追逐问题（方向相同）

Denote Alex's speed as r_A and Bob's speed as r_B with $r_A > r_B$. Let the distance between them be *d*, as in the following figure.

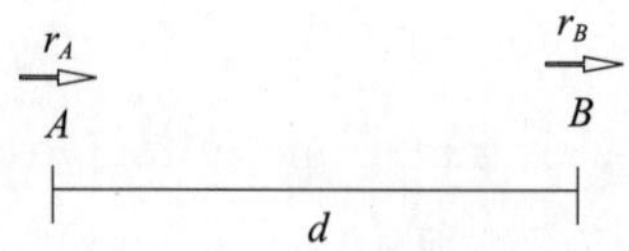

In order to simplify this problem, one may think that Bob is not moving, and that Alex walks in a relative speed of $(r_A - r_B)$ to cover a distance of *d* in time *t*. Then we have $d=(r_A - r_B)t$.

Therefore, the time needed for Alex to catch Bob is $t=\dfrac{d}{r_A - r_B}$.

设亚历克斯的速度为 r_A，鲍勃的速度为 r_B，并且 $r_A > r_B$。设他们之间的距离为 *d*，如图所示。

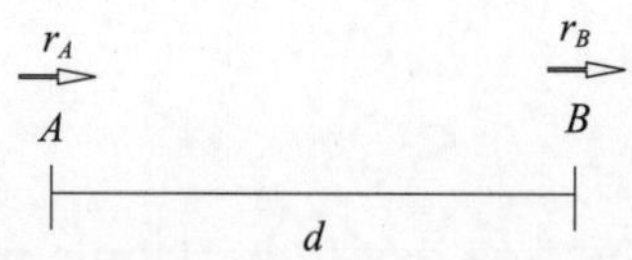

为了简化问题，可以假设鲍勃不动，而亚历克斯以相对速度（$r_A - r_B$）移动，在时间 *t* 内走完距离 *d*。则有 $d=(r_A - r_B)t$。

因此，亚历克斯追上鲍勃所需的时间为 $t=\dfrac{d}{r_A - r_B}$。

5.3.3 Two People Walking in the Opposite Directions (towards each other) 相遇问题（方向相反）

Alex's speed is r_A and Bob's speed is r_B and distance between them is *d*, as in the following figure. This time, one may think that Bob is not moving and Alex walks in a relative speed of $(r_A + r_B)$ to cover a distance of *d* with time *t*. Thus, $d=(r_A + r_B)t$, and the time needed for Alex to meet Bob is: $t=\dfrac{d}{r_A + r_B}$.

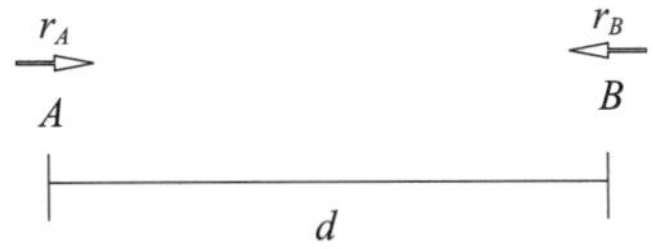

亚历克斯的速度为 r_A，鲍勃的速度为 r_B，且他们之间的距离为 d，如图所示。我们可以假设鲍勃不动，而亚历克斯以相对速度（$r_A + r_B$）行走完距离 d。所以 $d = (r_A + r_B)t$，而亚历克斯和鲍勃相遇所需的时间为 $t = \dfrac{d}{r_A + r_B}$。

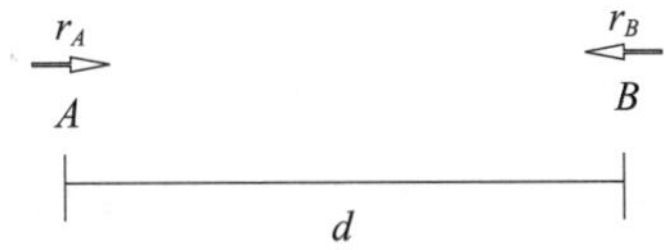

Example 11

As Amy is riding her bicycle on a long straight road, she spots Mike running in the same direction $\frac{2}{3}$ mile in front of her. After she passes him, she can see him in her rear mirror until he is $\frac{2}{3}$ mile behind her. Amy rides at a constant rate of 14 miles per hour, and Mike runs at a constant rate of 10 miles per hour. For how many minutes can Amy see Mike?

例题 11

当艾米在一条长直路上骑自行车时，她发现麦克在前方 $\frac{2}{3}$ 英里处朝同一方向跑步。在她超过他之后，她可以在后视镜中看到他，直到他距离她 $\frac{2}{3}$ 英里就在后视镜中消失了。已知艾米以 14 英里每小时的速度骑行，而麦克以 10 英里每小时的速度跑步。这段路程艾米看到麦克的时间有多少分钟？

(A) 12　　(B) 18　　(C) 20　　(D) 24　　(E) 30

Solution

Because they are both moving in the same direction, Amy is riding at a speed of $14 - 10 = 4$ mph relative to Mike. It's the same as Mike was still and Amy was approaching him in the speed of 4 mph.

It takes Amy $\frac{2}{3}\cdot\frac{1\text{ hour}}{4\text{ miles}}=\frac{1}{6}$ hour to reach Mike, and then the same amount of time to be $\frac{2}{3}$ miles ahead of him. The total amount of time is $2\cdot\frac{1}{6}$ hour$\cdot\frac{60\text{ minutes}}{1\text{ hour}}=20$ minutes. The answer is (C).

解析

这是一个追逐问题，艾米与麦克的速度差是 $14-10=4$ 英里每小时。也就是说，假设麦克站在原地没有移动，而艾米以 4 英里每小时的速度前进，那么她需要花 $\frac{2}{3}\times\frac{1}{4}=\frac{1}{6}$ 小时才能追上他，然后再花同样的时间超过他 $\frac{2}{3}$ 英里。总共花费的时间是 $2\times\frac{1}{6}\times\frac{60}{1}=20$ 分钟，故选 (C)。

Example 12

Betty begins to walk from her house toward her friend Elsa's house. At the same time, Elsa begins to ride her bicycle toward Betty's house. They each maintain a constant speed, and Elsa rides 4 times as fast as Bella walks. The distance between their houses is 3 miles, which is 15840 feet, and Betty covers 2 feet with each step. How many steps will Betty take by the time she meets Elsa?

例题 12

贝蒂从她家开始步行前往她的朋友埃尔莎的家。同时，埃尔莎骑自行车向贝蒂的家驶去。她们各自保持恒定的速度，且埃尔莎的骑行速度是贝蒂步行速度的 4 倍。她们两家的距离是 3 英里，约为 15850 英尺，而贝蒂每一步走 2 英尺。当贝蒂与埃尔莎相遇时，贝蒂一共走了多少步?

(A) 816　(B) 1280　(C) 1468　(D) 1585　(E) 2146

Solution

Every 10 feet Betty moves, Elsa does 40 feet, which means a total of 50 feet. Treat this as a section. Then they need to travel $15850 \div 50 = 317$ sections to cover the entire 3 miles. Since Betty moves 10 feet 317 times, this means that she travels 3170 feet in total. And since she walks 2 feet each step, the number of her steps is $3170 \div 2 = \text{(D)}1585$.

解　析

每当贝蒂走 10 英尺时，埃尔莎会走 40 英尺，即她们一共会移动 50 英尺。将此视为一个单元。他们大约一共需要移动 $15850 \div 50 = 317$ 个单元才能走完 3 英里。因为贝蒂移动了 317 次 10 英尺，所以她的移动距离为 3170 英尺。由于贝蒂每步走 2 英尺，因此她总共要走 $3170 \div 2 = 1585$ 步，故选 (D)。

Example 13

Cassie leaves Town A at 10:00 AM heading for Town B on her bike. She bikes at a constant rate of 14 miles per hour. Brian leaves Town B at 10:30 AM heading for Town A on his bike. He bikes at a constant rate of 18 miles per hour. They both bike on the same 55-mile route between Town A and Town B. At what time do they meet?

例题 13

卡茜在早上 10:00 离开 A 城，骑自行车前往 B 城。她的骑行速度是 14 英里每小时。布莱恩在早上 10:30 从 B 城出发，骑自行车前往 A 城，骑行速度是 18 英里每小时。他们都沿着 A 城和 B 城之间 55 英里长的同一路线骑行。问他们在几点钟会面?

(A) 11:15　　(B) 11:30　　(C) 11:45　　(D) 12:00　　(E) 12:30

Solution

Since Cassie leaves $\frac{1}{2}$ an hour earlier than Brian, Cassie has biked $14 \times \frac{1}{2} = 7$ miles when Brian leaves Town B, so the distance between them is $55 - 7 = 48$ miles at that time. After each hour, they get $14 + 18 = 32$ miles closer to each other. After $\frac{48 \text{ miles}}{32 \text{ mph}} = 1.5$ hours they will meet, at (D)12:00.

解 析

因为卡茜比布莱恩早出发半小时，那么当布莱恩出发时，卡茜已经骑行 $14\times\frac{1}{2}=7$ 英里，他们之间的距离是 $55-7=48$ 英里。从这里开始运用相遇问题的解法。每小时他们会相互靠近 $14+18=32$ 英里，因此 $\frac{48}{32}=1.5$，即 1.5 小时后他们会相遇，时间是从上午 10:30 开始的 1.5 小时后，也就是 12:00，故选 (D)。

PRACTICE 练习

Problem 1

Karl bought six folders from a supermarket at a cost of \$1.50 each. The supermarket had a 30%-off sale the following day. How much could Karl have saved on the purchase by waiting a day?

问题 1

卡尔从超市购买了 6 个文件夹，每个文件夹的价格是 1.50 美元。第二天，超市降价 30% 进行销售。卡尔如果等一天再买可以节省多少美元？

(A) 1.50　　(B) 2.40　　(C) 2.70　　(D) 2.75　　(E) 6.30

Problem 2

Two 800 mL pitchers contain orange juice. One pitcher is $\frac{1}{4}$ full and the other pitcher is $\frac{3}{5}$ full. Water is added to fill each pitcher completely, then both pitchers are poured into one large container. What fraction of the mixture in the large container is orange juice?

问题 2

两个 800 毫升的水壶里装有橙汁。一个水壶装了 $\frac{1}{4}$，另一个水壶装了 $\frac{3}{5}$。现在往两个水壶里继续装水以使每个水壶完全装满，然后将两个水壶里的混合液体倒入一个大容器中。大容器混合液体中的橙汁所占的比例是多少？

(A) $\frac{1}{8}$　　(B) $\frac{5}{16}$　　(C) $\frac{9}{20}$　　(D) $\frac{17}{40}$　　(E) $\frac{11}{20}$

Problem 3

A company sells detergent in three different sized boxes: small (S), medium (M) and large (L). The medium size costs 30% more than the small size and contains 30% less detergent than the large size. The large size contains twice as much detergent as the small size and costs 40% more than the medium size. Rank the three sizes from best to worst buy.

问题 3

一家公司出售 3 种不同大小的洗衣粉盒：小号（S）、中号（M）和大号（L）。中号的价格比小号贵 30%，且所盛的洗衣粉比大号少 30%。大号所盛的洗衣粉是小号的两倍，价格比中号贵 40%。请将这 3 种尺寸的洗衣粉盒按性价比从低到高排序。

(A) SLM　　(B) SML　　(C) LMS　　(D) LSM　　(E) MLS

Problem 4

Ryan got 60% of the problems correct on a 25-problem test, 70% on a 40-problem test, and 85% on a 20-problem test. About what percent of all the problems did Ryan answer correctly?

问题 4

瑞安在一场包含 25 道题的测试中答对了 60% 的题目，在一场包含 40 道题的测试中答对了 70% 的题目，在一场包含 20 道题的测试中答对了 85% 的题目。瑞安答对题目的总数量占总题目数量的百分比是多少？

(A) 64　　(B) 70　　(C) 74　　(D) 76　　(E) 78

Problem 5

A recipe that makes 4 servings of hot chocolate requires 3 squares of chocolate, $\frac{1}{3}$ cup sugar, 1.5 cup water and 3 cups milk. Jordan has 7 squares of chocolate, 1.5 cups of sugar, lots of water and 8 cups of milk. If she maintains the same ratio of ingredients, what is the greatest number of servings of hot chocolate she can make?

问题 5

一个配方可以制作 4 份热巧克力饮料，所需的材料为 3 块巧克力、$\frac{1}{3}$ 杯糖、1.5 杯水和 3 杯牛奶。乔丹有 7 块巧克力、1.5 杯糖、很多水和 8 杯牛奶。如果她保持相同的配料比例，那么最多可以制作多少份热巧克力饮料？

(A) 5　　(B) 6　　(C) 7　　(D) 8　　(E) 9

Problem 6

24 friends met for dinner at Oscar Oyster House, and each ordered one meal. The portions were so large, there was enough food for 32 people. If they shared, how many meals should they have ordered to have just enough food for the 24 of them?

问题 6

24 个朋友在奥斯卡生蚝餐厅共进晚餐，每人点了一份餐。由于分量非常大，足够 32 个人吃。如果可以分享的话，他们应该点多少份餐正好足够 24 个人吃？

(A) 18　　(B) 20　　(C) 21　　(D) 22　　(E) 27

Problem 7

The sales tax rate in a city is 9%. During a sale, the price of a coat is discounted 30% from its $120.00 price. Two clerks, Jack and Jill, calculate the bill independently. Jack rings up $120.00 and adds 9% sales tax, then subtracts 30% from this total. Jill rings up $120.00, subtracts 30% of the price, then adds 9% of the discounted price for sales tax. What is Jack's total minus Jill's total?

问题 7

一个城市的销售税率为 9%。在一次促销中，一件外套的原价是 120 美元，降价 30%。两位店员杰克和吉尔独立计算账单。杰克先输入 120 美元并加上 9% 的销售税，然后从这个总额中减去 30%。吉尔先输入 120 美元，减去价格的 30%，然后再加上折扣后价格的 9% 作为销售税。问杰克的总额减去吉尔的总额是多少？

(A) –$2.46　　(B) –$1.08　　(C) $0　　(D) $1.08　　(E) $2.46

Problem 8

There are 280 students at Calif Middle School, where the ratio of boys to girls is 3 : 4. There are 210 students at Wins Middle School, where the ratio of boys to girls is 4 : 3. The two schools hold a dance and all students from both schools attend. What fraction of the students at the dance are girls?

问题 8

加州中学有 280 名学生，男孩与女孩的比例为 3 : 4。温斯中学有 210 名学生，男孩与女孩的比例为 4 : 3。这两所中学举办了一场舞会，所有学生都参加。参加舞会的学生中女孩的比例是多少？

(A) $\frac{29}{70}$ (B) $\frac{23}{49}$ (C) $\frac{25}{49}$ (D) $\frac{37}{70}$ (E) $\frac{27}{49}$

Problem 9

Miki has a dozen oranges of the same size and a dozen pears of the same size. Miki uses her juicer to extract 12 ounces of pear juice from 4 pears and 16 ounces of orange juice from 3 oranges. She makes a pear-orange juice blend from an equal number of pears and oranges. What percent of the blend is pear juice?

问题 9

米琪有一打相同大小的橙子和一打相同大小的梨。米琪使用榨汁机从 4 个梨中提取了 12 盎司梨汁，从 3 个橙子中提取了 16 盎司橙汁。她用相同数量的梨和橙子制作了梨汁、橙汁混合饮料。那么，在这种混合饮料中梨汁占多少百分比？

(A) 30 (B) 36 (C) 40 (D) 42 (E) 48

Problem 10

Handy Aaron helped a neighbor $1\frac{1}{6}$ hours on Monday, 48 minutes on Tuesday, from 9:23 to 11:55 on Saturday morning, and a half-hour on Sunday. He is paid \$3 per hour. How much did he earn for the week?

问题 10

手巧的亚伦在星期一帮助邻居 $1\frac{1}{6}$ 小时，在星期二帮助了 48 分钟，在星期六早上从 9:23 帮助到 11:55，星期日则帮助了半小时。他的工资是每小时 3 美元。问他这一周赚了多少钱？

(A) \$24.5 (B) \$27 (C) \$28 (D) \$31.5 (E) \$35

Problem 11

A number of students from Fibonacci High School are taking part in a community service project. The ratio of 12th-graders to 10th-graders is 7 : 4, and the ratio of 12th-graders to 11th-graders is 9 : 5. What is the smallest number of students that could be participating in the project?

问题 11

一些来自斐波那契高中的学生正在参加一个社区服务项目。12 年级学生与 10 年级学生的比例为 7 : 4，而 12 年级学生与 11 年级学生的比例为 9 : 5。参与该项目的学生最少有多少人？

(A) 63 (B) 86 (C) 105 (D) 134 (E)181

Problem 12

Of the 200 balls in a large bag, 70% are red and the rest are blue. How many of the blue balls must be removed from the bag so that 80% of the remaining balls are red?

问题 12

在一个大袋子里有 200 个球，其中 70% 是红色的，其余是蓝色的。需要从袋子中移除多少个蓝球，才能使剩下的球中 80% 是红球？

(A) 15 (B) 20 (C) 25 (D) 30 (E) 40

Problem 13

The clock in Siri's car is not accurate and gains time at a constant rate. One day, when Siri begins shopping, both his car clock and his accurate watch show 10:00 AM. When he finishes shopping, his watch reads 11:00 AM, but the car clock reads 11:10 AM. Later that day, Siri loses his watch. When he checks the car clock, it shows 5:00 PM. What is the actual time?

问题 13

希瑞车上的时钟不准确，但以恒定的速度走。某一天，他开始购物时注意到车上的时钟和他的手表都显示上午 10:00。当他购物结束时，他的手表显示上午 11:00，而车上的时钟显示上午 11:10。当天稍晚时，希瑞把他的手表弄丢了。他查看车上的时间，发现显示下午 5:00。那么实际时间是什么时候？

(A) 4:00 (B) 4:20 (C) 4:30 (D) 4:50 (E) 5:40

Problem 14

Ted's grandfather used his treadmill on 3 days this week. He went 3 miles each day. On Monday he jogged at a speed of 6 miles per hour. He walked at the rate of 4 miles per hour on Wednesday and at 5 miles per hour on Friday. If Grandfather had always walked at 5 miles per hour, he would have spent less time on the treadmill. How many minutes less?

问题 14

泰德的祖父本周使用了 3 天跑步机。他每天走 3 英里。周一，他以 6 英里每小时的速度慢跑；周三，他以 4 英里每小时的速度走路；周五，他以 5 英里每小时的速度快走。如果祖父一直以 5 英里每小时的速度走路，他会在跑步机上少花费多少分钟？

(A) 2 (B) 3 (C) 4 (D) 5 (E) 6

Problem 15

Karl's car uses a gallon of gas every 40 miles, and his gas tank holds 16 gallons when it is full. One day, Karl started with a full tank of gas, drove 400 miles, bought 9 gallons of gas, and continued driving to his destination. When he arrived, his gas tank was $\frac{1}{4}$ full. How many miles did Karl drive that day?

问题 15

卡尔的车每 40 英里消耗 1 加仑汽油，他的油箱满时可容纳 16 加仑汽油。一天，卡尔驾驶满油箱汽车出发，行驶了 400 英里，购买了 9 加仑汽油，然后继续开往目的地。当他到达时，油箱只剩下 $\frac{1}{4}$ 箱汽油。问卡尔那天总共开车行驶了多少英里？

(A) 520 (B) 600 (C) 680 (D) 720 (E) 840

Problem 16

Jeremy's father drives him to school in rush hour traffic in 40 minutes. One day there is no traffic, so his father can drive him 12 miles per hour faster and gets him to school in 24 minutes. How far in miles is it to school?

问题 16

杰里米的父亲在交通拥堵时段用 40 分钟把他送到学校。在交通顺畅时段，他的父亲可以以 12 英里每小时的速度行驶，并在 24 分钟内把他送到学校。问从家到学校的距离是多少英里？

(A) 10 (B) 12 (C) 16 (D) 480 (E) 720

Problem 17

Niki usually leaves her cell phone on. If her cell phone is on but she is not actually using it, the battery will last for 48 hours. If she is using it constantly, the battery will last for only 6 hours. Since the last recharge, her phone has been on 18 hours, and during that time she has used it for 2 hours. If she doesn't talk any more but leaves the phone on, how many more hours will the battery last?

问题 17

尼基通常会让她的手机保持开启。如果她的手机开着但没有实际使用，电池能持续 48 小时。如果她一直在使用手机，电池只能持续 6 小时。自上次充满电开始，她的手机已经保持开启了 18 个小时，在这段时间内她使用了 2 个小时。如果她不再使用但保持手机开启，电池还能持续多少个小时？

(A) 10　　(B) 12　　(C) 14　　(D) 16　　(E) 20

Problem 18

Joe had walked half way from home to school when he realized he was late. He ran the rest of the way to school. He ran 4 times as fast as he walked. Joe took 16 minutes to walk half way to school. How many minutes did it take Joe to get from home to school?

问题 18

乔从家到学校，走了一半路程时意识到自己迟到了。他开始跑步去学校。他跑步的速度是走路速度的 4 倍。乔用 16 分钟走了一半路程。乔从家到学校一共用了多少分钟？

(A) 18　　(B) 20　　(C) 24.5　　(D) 26　　(E) 28

Problem 19

Town A and B are 120 miles apart. Bonnie drove from A to her daughter's house in B, averaging 90 miles per hour. Leaving the car with her daughter, Bonnie rode a bus back to A along the same route and averaged 60 miles per hour on the return trip. What was the average speed for the round trip in miles per hour?

问题 19

A 城和 B 城相距 120 英里。邦妮从奥斯丁开车到她的女儿在 B 城的家，平均速度为 90 英里每小时。将车留给女儿后，邦妮乘坐公交车沿同一路线返回 A 城，回程的平均速度为 60 英里每小时。这次往返旅行的平均速度是多少英里每小时？

(A) 70　　(B) 72　　(C) 75　　(D) 78　　(E) 80

Problem 20

Annie and Bonnie are running laps around a 200-meter oval track. They started together, but Annie has pulled ahead because she runs 20% faster than Bonnie. How many laps will Annie have run when she first passes Bonnie?

问题 20

安妮和邦妮正在一个 200 米长的椭圆形跑道上跑步。她们一起开始跑，安妮领先，因为她的速度比邦妮快 20%。问当安妮第一次超过邦妮时，她跑了多少圈？

(A) 2 (B) 3 (C) 4 (D) 5 (E) 6

Chapter 6 Equations and Inequalities

第 6 章 方程与不等式

6.1 Basic Knowledge 基础内容

6.1.1 Fundamental Law of Fractions 分数的基本法则

For any fraction $\frac{a}{b}$ and any number $c \neq 0$, $\frac{a}{b}=\frac{a\times c}{b\times c}$.

对任意分数 $\frac{a}{b}$，以及任意不等于 0 的数 c，都满足 $\frac{a}{b}=\frac{a\times c}{b\times c}$。

Example 1

What is the value of the product?

例题 1

下面这个乘积的值是多少？

$$\left(1+\frac{1}{1}\right)\times\left(1+\frac{1}{4}\right)\times\left(1+\frac{1}{10}\right)\times\left(1+\frac{1}{22}\right)\times\left(1+\frac{1}{46}\right)=?$$

(A) $\frac{1}{2}$ (B) $\frac{3}{2}$ (C) 2 (D) $\frac{47}{16}$ (E) $\frac{47}{8}$

Solution

By adding up the numbers in each of the 5 parentheses, we have:

$$\frac{2}{1}\times\frac{5}{4}\times\frac{11}{10}\times\frac{23}{22}\times\frac{47}{46}$$

Most of the terms in the numerator cancel out diagonally, leaving four 2s in the denominator. Eventually, we are left with $\frac{47}{2^4}$ which is equivalent to $\frac{47}{16}$. Thus the answer would be (D).

解 析

将 5 个括号中的数字相加可得

$$\frac{2}{1}\times\frac{5}{4}\times\frac{11}{10}\times\frac{23}{22}\times\frac{47}{46}$$

大部分分子上的数字与分母上的数字可以相互抵消，分子只剩下 47。同样地，分母上还剩下 4 个 2。最终得到的值是 $\frac{47}{2^4}$，即 $\frac{47}{16}$。因此答案是 (D)。

6.1.2 Division of Fractions 分数的除法

To divide by a fraction, we simply multiply by its reciprocal.

要除以一个分数，即乘以该分数的倒数。

$$\frac{a}{b}\div\frac{c}{d}=\frac{a}{b}\times\frac{d}{c}=\frac{ad}{bc}$$

6.1.3 Power Rules of Exponents 指数的幂法则

$$a^m\times a^n=a^{m+n}\Leftrightarrow a^{m+n}=a^m\times a^n$$

$$(a^m)^n=a^{mn}$$

$$(ab)^n=a^nb^n$$

$$\frac{a^m}{b^m}=\left(\frac{a}{b}\right)^m\Leftrightarrow\left(\frac{a}{b}\right)^m=\frac{a^m}{b^m}$$

Example 2

If $x✪y=(x^y)^x$, what is the units digit of 4✪8?

例题 2

如果 $x✪y=(x^y)^x$，那么 4✪8 的个位数字是多少？

(A) 6 (B) 8 (C) 2 (D) 4 (E) 0

Solution

According to the power rules of exponents, we get:

$$4✪8=(4^8)^4=4^{32}$$

The pattern for the last digit of 4^n is 4, 6, 4, 6, etc.

When the exponent 32 is divided by 2, the remainder is 0. Therefore, the last digit of 4^{32} is the same as the last digit of 4^2. So the last digit is 6. The correct answer is (A).

解 析

根据指数的幂法则可得

$$4✪8=(4^8)^4=4^{32}$$

4^n 的最后一位数字的规律是 4，6，4，6，以此类推。

可以看出，2 个数字一循环，当指数 32 除以 2 时，余数为 0。因此，4^{32} 的最后一位数字与 4^2 的最后一位数字相同，为 6。所以正确答案是 (A)。

6.1.4 Properties of Radicals 根式的性质

$$\sqrt{ab}=\sqrt{a}\times\sqrt{b}\ (a>0,\ b>0)$$

$$\frac{\sqrt{a}}{\sqrt{b}}=\sqrt{\frac{a}{b}}\ (a>0,\ b>0)$$

$$\sqrt[n]{a^m}=\left(\sqrt[n]{a}\right)^m=a^{\frac{m}{n}}$$

Example 3

What is the value of the expression $\sqrt{81\sqrt{27\sqrt{9}}}$?

例题 3

$\sqrt{81\sqrt{27\sqrt{9}}}$ 这个表达式的值是多少？

(A) 9 (B) $9\sqrt{3}$ (C) 27 (D) 243 (E) 729

Solution

According to properties of radicals, we have:

$$\sqrt{81\sqrt{27\sqrt{9}}}=\sqrt{81\times\sqrt{27\times3}}=\sqrt{81\times\sqrt{81}}=\sqrt{81\times9}=\sqrt{729}=27$$

Hence the correct answer is (C).

解 析

根据根式的性质可得

$$\sqrt{81\sqrt{27\sqrt{9}}}=\sqrt{81\times\sqrt{27\times 3}}=\sqrt{81\times\sqrt{81}}=\sqrt{81\times 9}=\sqrt{729}=27$$

因此正确答案是 (C)。

Example 4

For real numbers a and b, define $a \diamond b = \sqrt{a^2+b^2}$. What is the value of $(5 \diamond 12) \diamond ((-5) \diamond (-12))$?

例题 4

对于实数 a 和 b，我们定义 $a \diamond b=\sqrt{a^2+b^2}$。问 $(5 \diamond 12) \diamond ((-5) \diamond (-12))$ 的值是多少？

(A) 0 (B) $\frac{13}{2}$ (C) 13 (D) $13\sqrt{2}$ (E) 22

Solution

It follows from the definition that

$$(5\diamond 12)\diamond((-5)\diamond(-12))=\sqrt{5^2+12^2}\diamond\sqrt{(-5)^2+(-12)^2}=13\diamond 13=\sqrt{13^2+13^2}=13\sqrt{2}$$

Hence, the correct answer is (D).

解 析

根据定义可得

$$(5\diamond 12)\diamond((-5)\diamond(-12))=\sqrt{5^2+12^2}\diamond\sqrt{(-5)^2+(-12)^2}=13\diamond 13=\sqrt{13^2+13^2}=13\sqrt{2}$$

所以正确答案是 (D)。

6.1.5 Properties of Absolute Value 绝对值的性质

$$|-x|=|x|$$
$$|x-y|=|y-x|$$
$$|xy|=|x|\cdot|y|,\quad \left|\frac{x}{y}\right|=\frac{|x|}{|y|}\quad (y\neq 0)$$

6.1.6 Square Binomial 二项展开式(二次方)和平方差

$$(x+y)^2 = x^2 + 2xy + y^2$$
$$(x-y)^2 = x^2 - 2xy + y^2$$
$$x^2 - y^2 = (x-y)(x+y)$$

6.2 Equation 方程

An equation is a mathematical sentence in which the equal sign "=" connects two algebraic expressions.

方程是一个数学表达式，其中等号"="将两个代数表达式连接起来。

6.2.1 The Following are Equations

以下几个例子都是方程:

$2^{10}x = 1024$, $\frac{1}{7}x = 0.\overline{142857}$, $3\pi = 5x - 6$, $A = \pi r^2$

Open sentence: An equation that contains one or more variables is called an open sentence. For example: $2x + 2 = 8$.

开语句: 包含一个或多个未知量的方程称为开语句。例如，$2x + 2 = 8$。

Solution: A solution is a value of the variable that makes the equation true (Other names: root, and zero).

解: 解是使方程成立的变量的值(又称根、零点)。

For the equation $x + 5 = 9$, $x = 4$ is the solution since $4 + 5 = 9$.

When the variable in the equation is replaced by a constant so that the equation becomes true, the equation has been solved.

- We will learn the following skills:

1. get rid of the denominators
2. add parentheses
3. remove the parentheses
4. isolate the variable
5. combine like terms

- We will learn how to solve the following equations:

1. one-variable linear equations

2. literal equations

3. quadratic equations

4. system of linear equations

对于方程 $x+5=9$，$x=4$ 是解，因为 $4+5=9$。

当方程中的变量被替换为一个常数，且使方程成立时，方程就被解出。

- 我们将学习以下技能：

1. 去分母

2. 加括号

3. 去括号

4. 移项

5. 合并同类项

- 我们将学习如何解以下方程：

1. 一元一次方程

2. 数字谜

3. 一元二次方程

4. 多元一次方程组

6.3 Solving One-Variable Linear Equations 解一元线性方程（一元一次方程）

One-variable: one unknown in the equation, such as in $3x = 9$. x is the only unknown.

Linear: the variable has the power of 1, such as in $3x = 9$. x has the power of 1 ($x = x^1$).

The simplest form of this kind of equations: $ax = b$ ($a \neq 0$), where both a and b are constant. x is the variable (unknown).

The solution is $x = \dfrac{b}{a}$ (divide both sides of the equation by a).

一元：方程中存在一个未知数，例如 $3x=9$，x 是唯一的未知数。

线性（一次）：变量的次方为 1，例如在 $3x=9$ 中，x 的次方为 1（$x=x^1$）。

这类方程的最简形式为 $ax = b$（$a \neq 0$），其中 a 和 b 均为常数。x 是变量（未知数）。

解为 $x = \dfrac{b}{a}$（方程两边同时除以 a）。

Example 5

Ralph went to the store and bought 14 pairs of socks for a total of \$26. Some of the socks he bought cost \$1 a pair, some of the socks he bought cost \$3 a pair, and some of the socks he bought cost \$4 a pair. If he bought at least one pair of each type, how many pairs of \$1 socks did Ralph buy?

例题 5

拉尔夫去商店买了 14 双袜子，总共花了 26 美元。他买的袜子有些每双 1 美元，有些每双 3 美元，还有些每双 4 美元。如果他每种类型至少买一双，那么拉尔夫买了多少双 1 美元的袜子？

(A) 3　　(B) 6　　(C) 7　　(D) 9　　(E) 11

Solution

So let there be x pairs of \$1 socks, y pairs of \$3 socks, z pairs of \$4 socks. We have $x+y+z=14, x+3y+4z=26$, and $x\geqslant 1, y\geqslant 1, z\geqslant 1$.

Now we subtract the second equation from the first equation to find $2y+3z=12$, and $y,z\geqslant 1$. Rearranging to $z=\dfrac{12-2y}{3}$, we can observe that y is a multiple of 3 and $2y$ is a multiple of 6. Since $0<2y<12$, we must have $2y=6$. Therefore, $y=3$, and it follows that $z=2$ and $x=14-y-z=14-3-2=9$. The correct answer is (D).

解　析

设有 x 双 1 美元的袜子，y 双 3 美元的袜子，z 双 4 美元的袜子。可得 $x+y+z=14, x+3y+4z=26$，其中 $x\geqslant 1, y\geqslant 1, z\geqslant 1$。

二式减去一式，可得 $2y+3z=12$ 且 $y\geqslant 1, z\geqslant 1$。整理后得到 $z=\dfrac{12-2y}{3}$，因此 y 必须是3的倍数，即 $2y$ 是6的倍数。又因为 $0<2y<12$，可得 $2y=6$。因此，$y=3$，$z=2$，$x=14-y-z=14-3-2=9$。正确答案是 (D)。

Example 6

In a mathematics contest with ten problems, a student gains 4 points for a correct answer and loses 2 points for an incorrect answer. If Olivia answered every problem and her score was 16, how many correct answers did she have?

例题 6

在一个有 10 道题目的数学竞赛中，学生回答正确一题得 4 分，回答错误一题丢 2 分。如果奥利维亚回答了每一道题，她的得分为 16 分，那么她有多少道题回答正确？

(A) 4 (B) 5 (C) 6 (D) 7 (E) 8

Solution

Let x be the number of problems she answers correctly. Then the number she answered incorrectly is $10-x$ because she answers 10 questions in total. Her score is equal to $4x-2(10-x)=16$. Solving the equation, we get $4x+2x=16+20$. Hence $6x=36$ and we get $x=6$. The correct answer is (C).

解 析

设 x 为她回答正确的题目数量。因此她回答错误的题目数量为 $10-x$，她的得分可以写为 $4x-2(10-x)=16$。解这个方程，得 $4x+2x=16+20$。整理得到 $6x=36$，解得 $x=6$。故正确答案是 (C)。

Example 7

Each of the points A, B, C, D, E, and F in the figure below represents a different digit from 2 to 7. Each of the five lines shown passes through some of these points. The digits along each line are added to produce five sums, one for each line. The total of the five sums is 59. What is the digit represented by B?

例题 7

在下图中，每个点 A、B、C、D、E 和 F 各代表一个从 2 到 7 的不同数字。

五条线中每条线都穿过一些点。将每条线上的数字相加，一共可以得到 5 个和。这 5 个和的总和是 59。请问 B 代表的数字是什么？

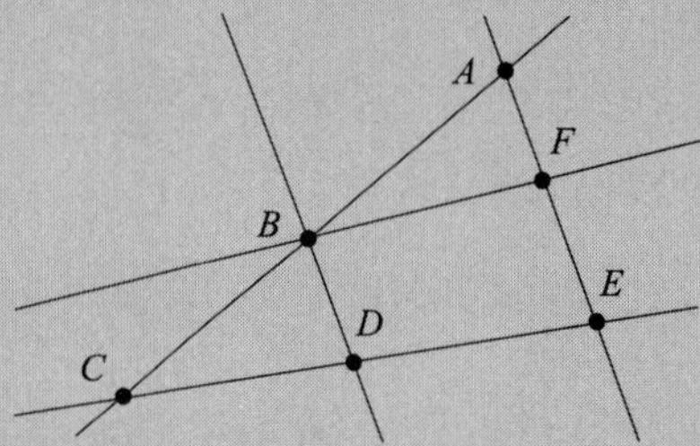

(A) 2　　(B) 3　　(C) 4　　(D) 5　　(E) 6

Solution

We can form the following expressions for the sum along each line:

$$A+B+C$$

$$A+E+F$$

$$C+D+E$$

$$B+D$$

$$B+F$$

Adding these together, we have $2A+3B+2C+2D+2E+2F = 59$, therefore $2(A+B+C+D+E+F)+B=59$. Since A, B, C, D, E, F are distinct integers between 2 and 7, we obtain $A+B+C+D+E+F=2+3+4+5+6+7=27$ (where the order doesn't matter as addition is commutative), so our equation simplifies to $54+B=59$. This means $B=5$. The correct answer is (D).

解　析

根据题意，将每条线的和写出以下算式：

$$A+B+C$$

$$A+E+F$$

$$C+D+E$$

$$B+D$$

$$B+F$$

将以上五个算式相加可得：$2A+3B+2C+2D+2E+2F=59$，即 $2(A+B+C+D+E+F)+$

$B=59$。由于 A、B、C、D、E 和 F 是 2 到 7 之间的 6 个不同的整数，因此：$A+B+C+D+E+F=2+3+4+5+6+7=27$（根据加法交换律，这里 A、B、C、D、E 和 F 的对应值并不重要），所以方程可以简化为 $54+B=59$，解得 $B=5$。所以正确答案是 (D)。

6.4 Solving Literal Equations 解文字方程

A literal equation is an equation that contains one or more letters. These letters are constants but are not fixed values.

General case: x is the variable in the equation.

Case Ⅰ. When $a \neq 0$, the equation has a unique solution: $x=\frac{b}{a}$.

Case Ⅱ. When $a=b=0$, the equation has infinite many solutions.

Case Ⅲ. When $a=0$ and $b \neq 0$, the equation has no solutions.

文字方程是包含一个或多个字母的方程。这些字母是常数，但没有固定取值。

一般情况下：x 是方程中的变量。

情况Ⅰ：当 $a \neq 0$ 时，方程有唯一解 $x=\frac{b}{a}$。

情况Ⅱ：当 $a=b=0$ 时，方程有无数个解。

情况Ⅲ：当 $a=0$ 且 $b \neq 0$ 时，方程没有解。

6.4.1 Quadratic Equations 二次方程的解

The following equation is called quadratic equation (the maximum power of x is x^2):

以下方程称为二次方程：（未知数 x 最大的幂次项为 x^2）

$$ax^2+bx+c=0 \ (a \neq 0)$$

6.4.2 Quadratic Formula 二次相关的公式

We will derive the quadratic formula below. The method used is called "completing the square" method. The method works for all quadratic equations.

$$ax^2+bx+c=0 \ (a \neq 0)$$

Since $a \neq 0$, we can divide both sides by a to get

$$x^2+\frac{b}{a}x+\frac{c}{a}=0$$

Now we complete the square:

$$x^2+2\times x\times\frac{b}{2a}+\left(\frac{b}{2a}\right)^2-\left(\frac{b}{2a}\right)^2+\frac{c}{a}=0$$

$$\uparrow\quad\uparrow\quad\uparrow\quad\uparrow\quad\uparrow$$

$$x^2+2\quad x\quad y\quad+\quad y^2\quad=(x+y)^2$$

$$x^2+2\times x\times\frac{b}{2a}+\left(\frac{b}{2a}\right)^2=\left(\frac{b}{2a}\right)^2-\frac{c}{a}$$

We can write the left side as a perfect square, and the right side as a single fraction:

$$\left(x+\frac{b}{2a}\right)^2=\frac{b^2-4ac}{4a^2}$$

Taking the square root of each side gives

$$x+\frac{b}{2a}=\pm\sqrt{\frac{b^2-4ac}{4a^2}}$$

Which implies

$$x_{1,2}=\frac{-b}{2a}\pm\sqrt{\frac{b^2-4ac}{4a^2}}=\frac{-b\pm\sqrt{b^2-4ac}}{2a}$$

i. e.

$$x_{1,2}=\frac{-b\pm\sqrt{b^2-4ac}}{2a}$$

我们将在下面推导二次方程相关的公式。所用的方法称为“配平方”法。该方法适用于所有二次方程。

$$ax^2+bx+c=0\ (a\neq 0)$$

由于$a\neq 0$，方程两边同时除以a得

$$x^2+\frac{b}{a}x+\frac{c}{a}=0$$

配方可得

$$x^2+2\times x\times\frac{b}{2a}+\left(\frac{b}{2a}\right)^2-\left(\frac{b}{2a}\right)^2+\frac{c}{a}=0$$

$$\uparrow\quad\uparrow\quad\uparrow\quad\uparrow\quad\uparrow$$

$$x^2+2\quad x\quad y\quad+\quad y^2\quad=(x+y)^2.$$

$$x^2+2\times x\times\frac{b}{2a}+\left(\frac{b}{2a}\right)^2=\left(\frac{b}{2a}\right)^2-\frac{c}{a}$$

我们可以将左侧写成一个完全平方数，而右侧写成一个分数：

$$\left(x+\frac{b}{2a}\right)^2=\frac{b^2-4ac}{4a^2}$$

方程左右两边开平方得到

$$x+\frac{b}{2a}=\pm\sqrt{\frac{b^2-4ac}{4a^2}}$$

解出 x 得

$$x_{1,2}=\frac{-b}{2a}\pm\sqrt{\frac{b^2-4ac}{4a^2}}=\frac{-b\pm\sqrt{b^2-4ac}}{2a}$$

化简可得

$$x_{1,2}=\frac{-b\pm\sqrt{b^2-4ac}}{2a}$$

Example 8

On the last day of school, Mrs. Susan gave her class stickers for their good performance throughout the term. She gave each boy as many stickers as there were boys in the class. She gave each girl as many stickers as there were girls in the class. She brought 400 pieces of stickers, and when she finished, she had six stickers left. There were two more boys than girls in her class. How many students were in her class?

例题 8

在学校的最后一天，苏珊老师为了奖励学生们一个学期来的优秀表现，给班级里的学生发了贴纸。她给每个男生的贴纸数量与班级中的男生数量相等。她给每个女生的贴纸数量与班级中的女生数量相等。她一共带来了 400 张贴纸，分完后，剩下 6 张。已知班级中的男生比女生多 2 人。问她的班级有多少名学生?

(A) 26 (B) 28 (C) 30 (D) 32 (E) 34

Solution

If there are x girls, then there are $x+2$ boys in the class. According to the question, Mrs. Susan gave each girl x stickers and each boy $x+2$ stickers. Therefore she gave away a total of $x^2+(x+2)^2$ stickers, which is also equivalent to $400-6=394$. The following equation can be made and simplified:

$$x^2+(x+2)^2=394$$
$$x^2+x^2+4x+4=394$$
$$2\cdot x^2+4x+4=394$$
$$2\cdot x^2+4x-390=0$$
$$x^2+2x-195=0$$
$$(x+15)(x-13)=0$$

The solution is $x=-15$ or 13. There cannot be a negative amount of students, therefore $x=13$. There are 13 girls and $13+2=15$ boys. In total there are $13+15=28$ students. The correct answer is (B).

解 析

设班级有 x 个女生，$x+2$ 个男生。根据题意，苏珊老师给每个女生 x 张贴纸，给每个男生 $x+2$ 张贴纸。因此，她总共发放的贴纸数量为 $x^2+(x+2)^2$，即 $400-6=394$ 张贴纸。可列出下面方程并化简：

$$x^2+(x+2)^2=394$$
$$x^2+x^2+4x+4=394$$
$$2\cdot x^2+4x+4=394$$
$$2\cdot x^2+4x-390=0$$
$$x^2+2x-195=0$$
$$(x+15)(x-13)=0$$

得到 $x=-15$ 或 $x=13$。由于学生人数不能为负，因此 $x=13$。于是，可得班级中有 13 个女孩和 $13+2=15$ 个男孩。总共是 $13+15=28$ 名学生。正确答案是 (B)。

6.4.3 Solving System of Linear Equations 解一元一次方程组

A group (two or more) of equations is called a system of equations. The solutions of a system of equations should satisfy all the equations. The most commonly used methods are substitution method and elimination method.

一组（两个或更多）方程称为方程组。方程组的解应为能同时满足方程组中每个方程的未知数的值。最常用的解方程组的方法有代入法和消元法。

Example 9

Solve the system of equations and find the value of $y-x$.

例题 9

解方程组并找出 $y-x$ 的值。

$$\begin{cases} \dfrac{x}{2}=\dfrac{y}{3} \\ \dfrac{y}{2}=\dfrac{z}{3} \\ x+y+z=38 \end{cases}$$

(A) 10 (B) 20 (C) 4 (D) 6 (E) 19

Solution

From the given equations, we have:

$$\frac{x}{4}=\frac{y}{6}=\frac{z}{9}=\frac{x+y+z}{4+6+9}=\frac{38}{19}=2$$

$$\frac{x}{4}=2$$

$$x=8$$

Substituting $x = 8$ into the equations, we get $y = 12$ and $z = 18$.

As such, $y-x=12-8=4$. The correct answer is (C).

解 析

从给定的方程组中可得

$$\frac{x}{4}=\frac{y}{6}=\frac{z}{9}=\frac{x+y+z}{4+6+9}=\frac{38}{19}=2$$

$$\frac{x}{4}=2$$

$$x=8$$

将 $x=8$ 代入方程，可得 $y=12$，$z=18$。

因此 $y-x=12-8=4$。正确答案是 (C)。

6.5 Linear Inequalities 一次不等式

6.5.1 Solve Inequalities 解不等式

To solve an inequality is to find the set of all values of the variable that make the inequality true.

This set of values is also known as the **solution set** of an inequality.

Two inequalities that have the same solution set are called **equivalent inequalities.**

解不等式就是找使不等式成立的所有变量可取值的集合。

这些值的集合也被称为不等式的**解集**。

具有相同解集的两个不等式称为**等价不等式**。

6.5.2 Sign of Inequality 不等式的符号

$a>b$	a is greater than b	a 大于 b
$a<b$	a is less than b	a 小于 b
$a\geqslant b$	a is greater than or equal to b	a 大于或等于 b
$a\leqslant b$	a is less than or equal to b	a 小于或等于 b
$a\neq b$	a is not equal to b	a 不等于 b

6.5.3 Rules of Solving Linear Inequalities 线性不等式求解的规则

(1) If $a>b$, then $a+c>b+c$;

(2) If $a>b$ and $c>0$, then $ac>bc$ (and $ca>cb$);

(3) If $a>b$ and $c<0$, then $ac<bc$ (and $ca<cb$).

(Note the change of the inequality symbol.)

(1) 如果 $a>b$，则 $a+c>b+c$；

(2) 如果 $a>b$ 且 $c>0$，则 $ac>bc$（和 $ca>cb$）；

(3) 如果 $a>b$ 且 $c<0$，则 $ac<bc$（和 $ca<cb$）。

（注意，这里的符号开口改变了方向。）

Example 10

For how many positive integer values of n are both $\frac{n}{3}$ and $2n$ three-digit whole numbers?

例题 10

有多少个正整数 n 可使 $\frac{n}{3}$ 和 $2n$ 都为三位数的整数？

(A) 33 (B) 66 (C) 67 (D) 166 (E) 167

Solution

For $\frac{n}{3}$ to be a three-digit whole number, n must be divisible by 3 and satisfy $\frac{n}{3} \geqslant 100$, so $n \geqslant 300$.

For $2n$ to be a three-digit whole number, n must satisfy $2n < 1000$ and $n < 500$.

Thus, $300 \leqslant n < 500$ and n must be divisible by 3.

There are $\frac{500}{3} - \frac{300}{3} + 1 = 67$ such numbers. The correct answer is (C).

解 析

为了使 $\frac{n}{3}$ 为三位数的整数，n 必须可被 3 整除且满足 $\frac{n}{3} \geqslant 100$，所以 $n \geqslant 300$。

为了使 $2n$ 为三位数的整数，n 必须满足 $2n < 1000$ 所以 $n < 500$。

因此，n 必须可被 3 整除且满足 $300 \leqslant n < 500$。

共有 $\frac{500}{3} - \frac{300}{3} + 1 = 67$ 个这样的数字。因此正确答案是 (C)。

PRACTICE 练习

Problem 1

Let the operations △ and □ be defined for all real numbers a and b as follows:

$$a \triangle b = a + 2b$$
$$a \square b = a + 5b$$

If 6 △ $(3y)$ = $(3y)$ □ 2, what is the value of y?

问题 1

对于所有实数 a 和 b，运算△ 和 □ 的定义如下：

$$a \triangle b = a + 2b$$
$$a \square b = a + 5b$$

如果 $6 \triangle (3y) = (3y) \square 2$，那么 y 的值是多少？

(A) $\frac{3}{2}$　　(B) 2　　(C) $\frac{4}{3}$　　(D) 1　　(E) 3

Problem 2

For the positive integer n, let <*n*> denote the sum of all the positive divisors of n with the exception of n itself. For example, <6> = 1 + 2 + 3 = 6 and <15> = 1 + 3 + 5 = 9. What is <<<20>>>?

问题 2

对于正整数 n，令 <*n*> 代表 n 的所有正因子的和但不包括 n 本身。例如，<6> = 1 + 2 + 3 = 6，而 <15> = 1 + 3 + 5 = 9。那么 <<<20>>> 的值是多少？

(A) 10　　(B) 12　　(C) 14　　(D) 96　　(E) 252

Problem 3

If $A \otimes B = \frac{A}{B} + \frac{B}{A}$, what is the value of $(6 \otimes 5) - (5 \otimes 6)$?

问题 3

如果 $A \otimes B = \frac{A}{B} + \frac{B}{A}$，那么 $(6 \otimes 5) - (5 \otimes 6)$ 的值是多少？

(A) $\frac{12}{5}$　　(B) $\frac{5}{3}$　　(C) 0　　(D) $\frac{6}{5}$　　(E) $\frac{5}{6}$

Problem 4

If $a \diamond b = \frac{1}{a} + \frac{1}{b}$, for what decimal value of a is $a \diamond 0.5 = 10$?

问题 4

如果 $a \diamond b = \frac{1}{a} + \frac{1}{b}$，那么下列哪个 a 值满足 $a \diamond 0.5 = 10$？

(A) $\frac{1}{8}$　　(B) $\frac{1}{5}$　　(C) 0　　(D) 1　　(E) 20

Problem 5

Harry and Terry are each told to calculate $12-(3+6)$. Harry gets the correct answer. Terry ignores the parentheses and calculates $12-3+6$. If Harry's answer is H and Terry's answer is T, what is $H-T$?

问题 5

哈里和特里分别来计算 $12-(3+6)$。哈里得到了正确答案。特里漏掉了括号，变成计算 $12-3+6$。如果哈里的答案是 H，特里的答案是 T，那么 $H-T$ 的值是多少？

(A) -12 (B) -5 (C) 0 (D) 5 (E) 12

Problem 6

In the AMC 8 contest students get 1 point if they answer a question correctly and 0 point otherwise. There are 25 questions. Billy answers 15 questions correctly, answers 6 questions incorrectly and doesn't answer the last 4. What is his score?

问题 6

在 AMC 8 比赛中，学生们如果答对一题得 1 分，答错或不答得 0 分。比赛共 25 道题。比利有 15 道题回答正确，6 道题回答错误，最后 4 道题没有回答。他的得分是多少？

(A) 5 (B) 9 (C) 10 (D) 15 (E) 19

Problem 7

Six lamp posts are equally spaced along one side of a straight road. The distance from the first lamp post to the fourth is 60 feet. What is the distance in feet between the first and last lamp posts?

问题 7

6 个路灯平均分布在一条笔直的路的一侧。第 1 个路灯到第 4 个路灯之间的距离是 60 英尺。那么第 1 个路灯和最后 1 个路灯之间的距离是多少英尺？

(A) 90 (B) 100 (C) 105 (D) 120 (E) 140

Problem 8

Anna and Bella are celebrating their birthdays together. Four years ago, when Bella turned 6 years old, she received a newborn puppy as a birthday present. Today the sum of the ages of the two children and the puppy is 28 years. How many years older than Bella is Anna?

问题 8

安娜和贝拉一起庆祝生日。4 年前，当贝拉满 6 岁时，她收到了一个刚出生的小狗作为生日礼物。今天，两个孩子和小狗的年龄总和是 28 岁。请问安娜比贝拉大多少岁？

(A) 3 (B) 4 (C) 5 (D) 6 (E) 7

Problem 9

A frog sits on a long line of lily pads in a pond. From any lily pad, the frog can jump 7 pads to the right or 4 pads to the left. What is the fewest number of jumps the frog must make to reach the lily pad located 1000 pads to the right of her starting point?

问题 9

一只青蛙蹲在池塘中长长的一排荷叶上。从任意荷叶出发，青蛙可以向右跳 7 片荷叶或向左跳 4 片荷叶。青蛙最少需要多少次跳跃才可以到达起始点右侧第 1000 片荷叶？

(A) 139 (B) 141 (C) 143 (D) 144 (E) 146

Problem 10

In a 25-question test, each correct answer is worth 4 points, each unanswered question is worth 1 point and each incorrect answer is worth 0 points. Which of the following scores is NOT possible?

问题 10

在一次测验中，共有 25 道题。答对 1 道得 4 分，没回答的问题得 1 分，答错 1 道题得 0 分。以下哪个分数是不可能的？

(A) 88 (B) 93 (C) 97 (D) 98 (E) 100

Problem 11

If 5! means $5\times4\times3\times2\times1$, what is the value of $\dfrac{7!}{5!}$?

问题 11

如果 5! 代表 $5\times4\times3\times2\times1$，那么 $\dfrac{7!}{5!}$ 的值是多少？

(A) 20 (B) 30 (C) 42 (D) 60 (E) 210

Problem 12

For all values *a*, *b*, *c*, and *d*, $\begin{vmatrix} a & c \\ d & b \end{vmatrix} = ab - cd$. If $\begin{vmatrix} 5 & x \\ -2 & 6 \end{vmatrix} = 8$. What is 2*x*?

问题 12

对于所有 a、b、c、和 d，定义 $\begin{vmatrix} a & c \\ d & b \end{vmatrix} = ab - cd$。如果 $\begin{vmatrix} 5 & x \\ -2 & 6 \end{vmatrix} = 8$，那么 $2x$ 的值是多少？

(A) -11 (B) 14 (C) -14 (D) 11 (E) -22

Problem 13

All the T-shirts in Maria's wardrobe are red, white, or yellow. Maria has half as many red shirts as white shirts and twice as many yellow shirts as white shirts. Which of the following could be the total number of T-shirts in Maria's wardrobe?

问题 13

在玛丽亚的衣柜中，短袖的颜色有红色、白色和黄色。其中，红色短袖的数量是白色短袖数量的一半，而黄色短袖的数量是白色短袖数量的两倍。玛丽亚衣柜中的短袖总数是多少？

(A) 24 (B) 25 (C) 26 (D) 27 (E) 28

Problem 14

Teacher Nancy has a total of 90 students in 4 classes.

✧ There are 1 more students in the 2nd class than in the 1st class.

✧ There are 2 more students in the 3rd class than in the 2nd class.

✧ There are 3 more students in the 4th class than in the 3rd class.

How many students are in the 4th class?

问题 14

南希老师的 4 个班级一共有 90 名学生。

✧ 第二个班里的学生比第一个班多 1 人。

✧ 第三个班里的学生比第二个班多 2 人。

✧ 第四个班里的学生比第三个班多 3 人。

问第四个班中有多少名学生？

(A) 20 (B) 21 (C) 23 (D) 24 (E) 26

Problem 15

Tony and Jack are both students in Ms. Lin's English class. Last night they each solved half of the problems in their homework alone and then solved the other half together. Tony had correct answers to only 80% of the problems he solved alone, but overall 88% of his answers were correct. Jack had correct answers to 90% of the problems he solved alone. What was Jack's overall percentage of correct answers?

问题 15

托尼和杰克都是林老师英语课的学生。昨晚，他们各自单独解决了一半的家庭作业题目，然后一起解决了另一半作业。托尼单独解决的问题中只有 80% 是正确的，但他的总体正确率为 88%。杰克单独解决的问题中有 90% 是正确的。杰克的总体正确率百分比是多少?

(A) 89　　(B) 92　　(C) 93　　(D) 96　　(E) 97

Problem 16

Before the mid-season break, the Twinkles had won 45% of their netball games. After the mid-season break, they won six more games and lost two, to finish the season having won half their games. How many games did the Twinkles play in all?

问题 16

在中期休赛前，闪耀队赢得了 45% 的英式篮球比赛。在中期休赛结束后，他们又赢得了 6 场比赛，输掉了 2 场，赛季结束时总共赢得了半数的比赛。那么闪耀队总共打了多少场比赛?

(A) 48　　(B) 50　　(C) 52　　(D) 54　　(E) 60

Problem 17

Matthew is a diligent worker. He is assigned the tasks of digging a tunnel of a certain distance. On the first day, Matthew dug $\frac{1}{5}$ of distance planned plus 12 meters more, and on the second day he dug $\frac{1}{4}$ of the remaining distance planned plus 15 meters. On the third day he dug $\frac{1}{3}$ of the remaining distance plus 18 meters. He then realized that there were only 62 meters left to dig, which he finished the next day. How many meters do he need to dig for the tunnel?

问题 17

马修是个勤劳的工人。他被安排去挖一条隧道。第一天，马修挖了总距离的 $\frac{1}{5}$ 加

12 米，第二天，他挖了剩余距离的 $\frac{1}{4}$ 加 15 米。第三天，他挖了剩余距离的 $\frac{1}{3}$ 加 18 米。然后他发现只剩下 62 米未挖，第四天他把剩余的距离挖完了。请问他一共挖了多少米的隧道？

(A) 120　　(B) 180　　(C) 240　　(D) 300　　(E) 360

Problem 18

A basketball league in the high school consists of two four-team divisions. Each team plays against every other team in its division X games. Each team plays against every team in the other division Y games with $X > 2Y$ and $Y > 4$. Each team plays a 76-game schedule. How many games does a team play within its own division?

问题 18

一个高中篮球联盟由两个拥有 4 支球队的联盟组成。每个球队在其联盟内与其他球队各比赛 X 场，与另一个联盟内的每支球队各比赛 Y 场，且需要满足 $X > 2Y$ 和 $Y > 4$。每支球队的比赛总数为 76 场。问一个球队在自己的联盟内总共需要比赛多少场？

(A) 36　　(B) 48　　(C) 54　　(D) 58　　(E) 60

Problem 19

Steph answered 15 questions out of 20 questions correctly in the first half of the math quiz, and 10 questions out of 10 questions correctly in the second half. Candace attempted 12 questions in the first half and 18 questions in the second. In each half, Steph scored a higher point than Candace. Surprisingly they ended with the same overall percentage of questions answered correctly. How many more questions did Candace score in the second half than in the first?

问题 19

斯蒂芬在第一场数学测验中总共答了 20 题，答对了 15 题，在第二场总共答了 10 题，他全部答对。坎迪斯在第一场测验中尝试了 12 题在第二场尝试了 18 题。在每场数学测验中，斯蒂芬的正确率都高于坎迪斯。令人惊讶的是他们的总正确率相同。问坎迪斯在第二场测验中比第一场多答对几道题？

(A) 7　　(B) 8　　(C) 9　　(D) 10　　(E) 11

Problem 20

Fifteen nails are arranged in order from the shortest to the longest according to their length (in mm) $a_1, a_2, a_3, \cdots, a_{15}$. The length of each nail must be integer and integers are equally spaced and have the property that

$$1 \leqslant a_1 \leqslant 10,\ 13 \leqslant a_2 \leqslant 20, \text{ and } 241 \leqslant a_{15} \leqslant 250$$

What is the sum of digits of a_{14}?

问题 20

15 个钉子按照长短（单位：mm）从短到长排列为 a_1，a_2，a_3，…，a_{15}。每个钉子的长度都是整数，且间隔的整数数值相等，并且满足以下条件：

$$1 \leqslant a_1 \leqslant 10,\ 13 \leqslant a_2 \leqslant 20,\quad 241 \leqslant a_{15} \leqslant 250$$

求 a_{14} 的各位数字之和。

(A) 7　　(B) 8　　(C) 9　　(D) 10　　(E) 11

Chapter 7 Set and Venn Diagrams

第 7 章 集合与韦恩图

7.1 Basic Knowledge 基础内容

7.1.1 Set 集合

A set is any well-defined collection of objects. Individual objects are called the elements or members of the set.

集合是任意一组良定义的对象汇总而成的。每个对象称为集合的元素或成员。

7.1.2 Subset 子集

A subset is a sub-collection of a set. All elements in a subset are in the original set. We denote set B as subset of A by the notation $B \subseteq A$.

子集是一个集合中的部分元素的汇总，子集中的元素均是原集合中的元素。我们用$B \subseteq A$来表示 B 是 A 的子集。

7.1.3 Proper Subset 真子集

If B is a subset of A and B is not equal to A, B is a proper subset of A, written as $B \subset A$.

如果 B 是 A 的子集并且 B 不等于 A，则 B 是 A 的真子集，记为 $B \subset A$。

7.1.4 Universal Set 全集

Universal set, denoted by U, is the set that contains all elements considered in a given discussion.

全集，通常用 U 表示，是在给定情境下包含所有被考虑元素的集合。

7.1.5 Empty Set 空集

A set with no element is called empty set, denoted by $\varnothing$. An empty set is a subset of any set, and it is also a proper subset of any nonempty set.

不含任何元素的集合称为空集，用符号 $\varnothing$ 表示。空集是任何集合的子集，也是任何非空集合的真子集。

7.1.6 Intersection of Sets 集合的交集

The intersection of sets A and B, written as $A \cap B$, is the set of all elements belonging to both A and B.

$$A \cap B \Rightarrow \text{common part of } A \text{ and } B$$

* One simple way to remember this: note $\cap$ is like the second letter in the word "and".

集合 A 和 B 的交集，记为 $A \cap B$，是 A 和 B 所有共有元素组成的集合。

$$A \cap B \Rightarrow A \text{ 和 } B \text{ 的公共部分}$$

* 一个简单的记忆方法：注意，符号 $\cap$ 像英文单词"and"的第二个字母"n"。

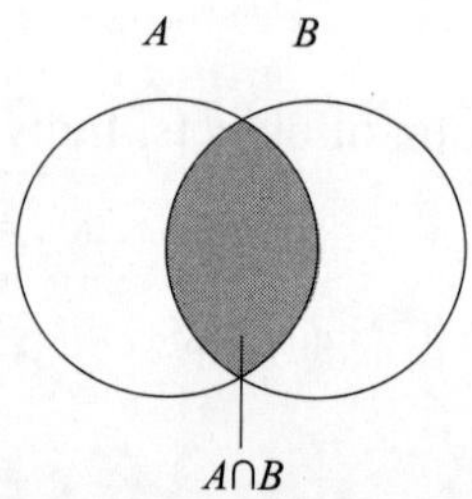

Example 1

Find $A \cap B$, if let $A = \{1, 2, 4, 5, 9\}$, and $B = \{1, 2, 3, 5, 7, 9\}$.

例题 1

已知 $A = \{1, 2, 4, 5, 9\}$，$B = \{1, 2, 3, 5, 7, 9\}$，求 $A \cap B$。

Solution

$A \cap B=\{1, 2, 5, 9\}$.

解　析

$A \cap B=\{1, 2, 5, 9\}$。

7.1.7 Union of Sets 集合的并集

The union of sets A and B, written as $A \cup B$, is the set of all elements belonging to either

A or B.

*Note the symbol $\cup$ is like the first letter in the word "union".

集合 A 和 B 的并集，记为 $A \cup B$，是属于 A 和 B 的所有元素组成的集合。

* 注意，符号 $\cup$ 像英语单词 "union" 的首字母 "u"。

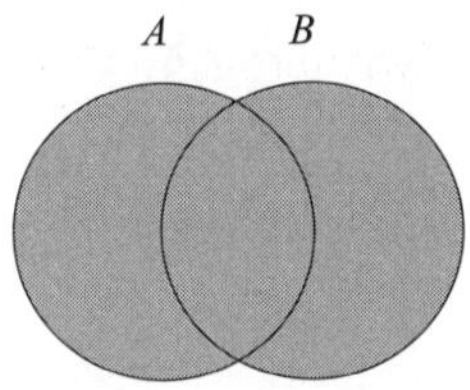

$A \cup B$

Example 2

Find $A \cup B$, if let $A = \{1, 2, 4, 5, 9\}$, and $B = \{1, 2, 3, 5, 7, 9\}$.

例题 2

已知 $A = \{1, 2, 4, 5, 9\}$，$B = \{1, 2, 3, 5, 7, 9\}$，求 $A \cup B$。

Solution

$A \cup B = \{1, 2, 3, 4, 5, 7, 9\}$.

解　析

$A \cup B = \{1, 2, 3, 4, 5, 7, 9\}$。

7.1.8 The Union Formula for Two Events 涉及两个事件的并集公式

$$n(A \cup B) = n(A) + n(B) - n(A \cap B)$$

Example 3

Among students in class 1A, $\frac{2}{5}$ of the students are wearing glasses, and $\frac{3}{4}$ of them are having pony tails. What is the minimum number of students in the class having both glasses and pony tails?

例题 3

1A 班级的学生中，有 $\frac{2}{5}$ 的人戴眼镜，有 $\frac{3}{4}$ 的人扎马尾辫。请问班级里至少有多少人同时戴眼镜和扎马尾辫？

(A) 3 (B) 5 (C) 8 (D) 15 (E) 20

Solution

Let x be the number of students wearing glasses and ponytails. Since the number of students wearing a pair of glasses or having ponytails must be whole numbers, the number of people in class 1A must be a multiple of lcm (4,5) =20. Since we are trying to find the minimum of x, we must use the smallest possible value for the number of students in class 1A. It follows that there are $\frac{2}{5}\times 20=8$ students wearing glasses and $\frac{3}{4}\times 20=15$ people having ponytails.

Let U be set of students in the class, A be the set of students wearing glasses and B be the set of students having ponytails, we can write:

$$n(U)\geqslant n(A\cup B)=n(A)+n(B)-n(A\cap B).$$

So, we have $20\geqslant 8+15-x$, and $x\geqslant 3$. Thus, (A) is the correct answer.

解　析

设 x 为同时戴眼镜和扎马尾辫的学生人数。由于戴眼镜或扎马尾辫的人数必须是整数，1A 班级里的学生数量必须是 4 和 5 的最小公倍数 20 的倍数。为了使 x 最小，必须取 1A 班级里学生人数的最小值。因此，有 $\frac{2}{5}\times 20=8$ 名学生戴眼镜，有 $\frac{3}{4}\times 20=15$ 名学生扎马尾辫。

设 U 为包含所有学生的集合，A 为戴眼镜的学生的集合，B 为扎着马尾辫的学生的集合，可得

$$n(U)\geqslant n(A\cup B)=n(A)+n(B)-n(A\cap B).$$

所以 $20\geqslant 8+15-x$，得 $x\geqslant 3$。因此，(A) 是正确答案。

Example 4

In the annual survey regarding the destinations for overseas learning experience this year, 198 students voted for two locations provided by the school with the following results: 149 voted in favor of Country A and 119 voted in favor of Country B. If there were exactly 29 students who voted against both destinations, how many students voted in favor of both destinations?

例题 4

在今年有关海外学习体验目的地的年度调查中，有 198 名学生对校方提供的两个目的地进行了投票，结果如下：149 人支持去国家 A，119 人支持去国家 B。如果有 29 人同时反对去这两个目的地，那么有多少人同时支持去这两个目的地？

(A) 49　　(B) 70　　(C) 79　　(D) 99　　(E) 149

Solution

Let A and B be the sets of students voting in favor of Country A and B, respectively. Then $A \cap B$ represents the set of students voting for both destinations. Since there are 29 students voting against both destinations, $n(A \cup B) = 198 - 29 = 169$.

By the Two Events Union Formula, we know:

$$n(A \cup B) = n(A) + n(B) - n(A \cap B)$$

Thus, we have $169 = 149 + 119 - n(A \cup B)$, and $n(A \cap B) = 99$.

As shown, (D) is the correct answer.

解　析

假设 A 和 B 分别是投票支持国家 A 和国家 B 的学生集合。那么 $A \cap B$ 代表同时支持两个目的地的学生的集合。由于有 29 个学生同时反对两个目的地，因此 $n(A \cup B) = 198 - 29 = 169$。

通过两个事件的并集公式，可得

$$n(A \cup B) = n(A) + n(B) - n(A \cap B)$$

所以 $169 = 149 + 119 - n(A \cap B)$，解得 $n(A \cap B) = 99$。

因此 (D) 是正确答案。

7.1.9 The Union Formula for Three Events 涉及三个事件的并集公式

The union of sets A, B, and C, written as $A \cup B \cup C$, is the set of all elements belonging to A, or B, or C.

集合A、B和C的并集，记为$A \cup B \cup C$，是属于A、B和C的所有元素构成的集合。

$$n(A \cup B \cup C) = n(A) + n(B) + n(C) - n(A \cap B) - n(B \cap C) - n(C \cap A) + n(A \cap B \cap C)$$

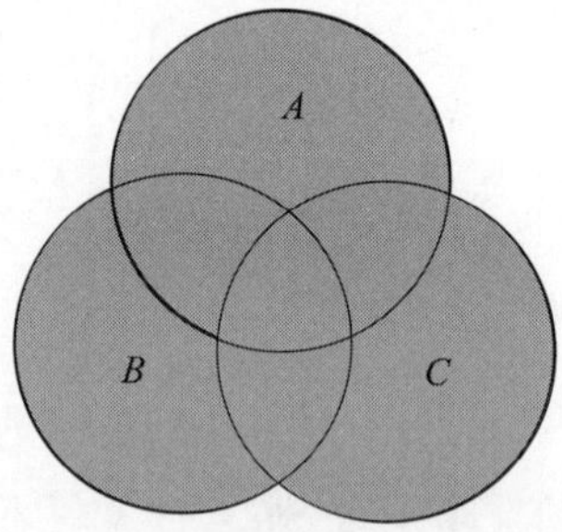

$A \cup B \cup C$

Example 5

Out of 22 students surveyed on the movie genre they like, 12 like Science Fiction, 5 like only Horror, and 6 liked Comedy. If 3 liked Science Fiction and Comedy, how many students did not like any of these genres?

例题 5

在一个针对 22 名学生进行的电影喜好类型调查中，12 人说他们喜欢科幻片，5 人说他们只喜欢恐怖片，6 人说他们喜欢喜剧片。如果 3 人又喜欢科幻片又喜欢喜剧片，那么有多少人同时不喜欢这三个电影类型？

(A) 2　(B) 9　(C) 3　(D) 5　(E) 6

Solution

Let A, B, and C be the sets of students favoring Science Fiction, only Horror, and Comedy, respectively. Let U be the universal set.

Then, we can write: $n(A) = 12$, $n(B) = 5$, $n(C) = 6$, $n(A \cap C) = 3$ and $n(U) = 22$.

By the formula of union for two events, we get that

$$n(A \cup C) = n(A) + n(C) - n(A \cap C) = 12 + 6 - 3 = 15$$

Since B is the set of students only favoring Horror, this implies that there is no intersection between the sets B and $A\cup C$.

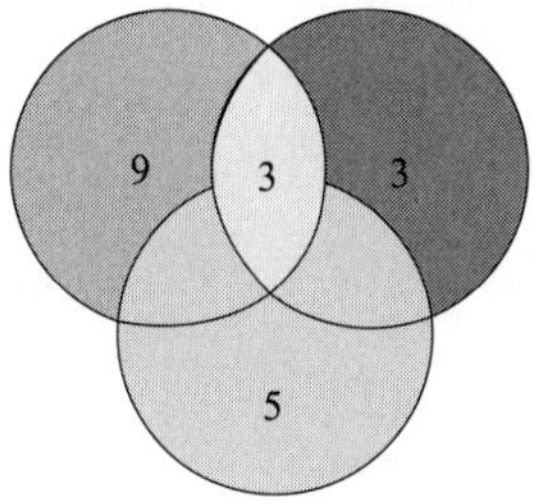

Therefore, the size of the union of A, B, and C can be found this way:

$$n(A\cup B\cup C)=n(A\cup C)+n(B)=15+5=20$$

which represents the number of students who likes one or more of these movie genres. Let the number of students who did not like any of these genres be x. Then we can obtain $x=22-n(A\cup B\cup C)$ and $x=2$. (A) is the correct answer.

解　析

设 A，B，C 分别为喜欢科幻片、只喜欢恐怖片和喜欢喜剧片的学生的集合。另设 U 为全集。

因此可得 $n(A)=12$，$n(B)=5$，$n(C)=6$，$n(A\cap C)=3$，$n(U)=22$。

根据两个事件的并集公式可知

$$n(A\cup C)=n(A)+n(C)-n(A\cap C)=12+6-3=15$$

由于 B 为只喜欢恐怖片的学生的集合，所以集合 B 与 $A\cup C$ 之间没有交集。

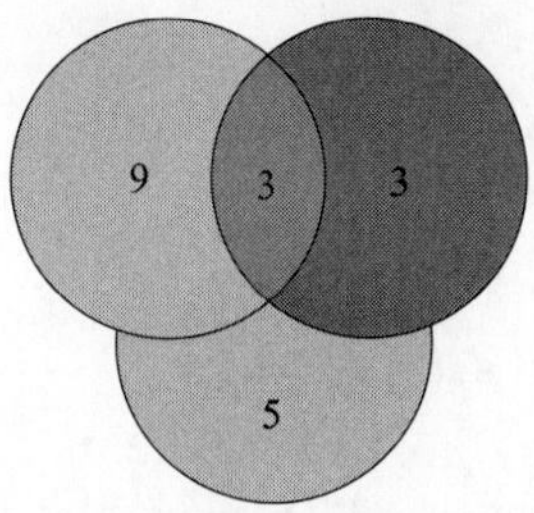

因此，集合 A，B 和 C 的并集可以通过如下方法求得：

$$n(A\cup B\cup C)=n(A\cup C)+n(B)=15+5=20$$

$n(A\cup B\cup C)$ 代表喜欢一种或多种电影类型的学生人数。设不喜欢这三个当中任何电影类型的学生人数为 x，可得 $x=22-n(A\cup B\cup C)$，因此 $x=2$。所以 (A) 是正确答案。

7.2 Calculation of the Number of Elements 元素个数的计算

The number of positive integers from 1 to n (not bigger than n) that are divisible by d is $\left\lfloor \frac{n}{d} \right\rfloor$, here $\lfloor x \rfloor$ is called the floor function. Whenever we see this notation, we take the greatest integer value that does not exceed x. For examples:

$$\lfloor 3.14 \rfloor = 3\ ,\ \lfloor 2.7 \rfloor = 2\ ,\text{ and } \lfloor 9.9 \rfloor = 9$$

从 1 到 n（不大于 n）的正整数中，能够被 d 整除的数量为 $\left\lfloor \frac{n}{d} \right\rfloor$，这里$\lfloor x \rfloor$称为向下取整函数。每当看到这个符号时，我们取小于等于 x 的最大的整数值。例如：

$$\lfloor 3.14 \rfloor = 3,\ \lfloor 2.7 \rfloor = 2,\ \lfloor 9.9 \rfloor = 9$$

Example 6

Consider the set $A = \{1, 2, 3, \cdots, 54\}$. How many are neither divisible by 3 nor 5?

例题 6

思考一下集合 $A = \{1，2，3，\cdots，54\}$ 中，有多少个数既不能被 3 整除也不能被 5 整除？

(A) 29 (B) 26 (C) 24 (D) 21 (E) 20

Solution

Let $A_3 \subset A$ be the set of those integers divisible by 3 and $A_5 \subset A$ be the set of those integers divisible by 5.

$$N(A_3 \cup A_5) = n(A_3) + n(A_5) - n(A_3 \cap A_5) = \left\lfloor \frac{54}{3} \right\rfloor + \left\lfloor \frac{54}{5} \right\rfloor - \left\lfloor \frac{54}{15} \right\rfloor = 18 + 10 - 3 = 25$$

Thus, the number of our target elements is $n(A) - n(A_3 \cup A_5) = 54 - 25 = 29$. The correct answer is (A).

解 析

设 $A_3 \subset A$ 为可被 3 整除的整数构成的集合，$A_5 \subset A$ 为可被 5 整除的整数所构成的集合。

$$N(A_3 \cup A_5) = n(A_3) + n(A_5) - n(A_3 \cap A_5) = \left\lfloor \frac{54}{3} \right\rfloor + \left\lfloor \frac{54}{5} \right\rfloor - \left\lfloor \frac{54}{15} \right\rfloor = 18 + 10 - 3 = 25$$

因此有 $n(A) - n(A_3 \cup A_5) = 54 - 25 = 29$ 个目标元素，正确答案是 (A)。

7.3 Problems Involving "At Least" or "At Most" 涉及"至少"或"最多"的问题

Example 7

Of the 100 students at the National High School, 70 are taking Physics and 75 are taking Chemistry. What is the fewest number of students at the National High School who could be taking both?

例题 7

在国立高中的 100 名学生中，70 人学习物理，75 人学习化学。问国立高中至少有多少名学生可能这两门课程都学习？

(A) 70 (B) 50 (C) 45 (D) 30 (E) 16

Solution

From the question, we know the universal set U satisfying $n(U) = 100$. Let A be the set of students taking Physics, B be the set of students taking Chemistry, x be the number of students taking neither class, and y be the number of students taking both. By the Two Events Union Formula, we know:

$$n(A \cup B) = n(A) + n(B) - n(A \cap B)$$

And we have $100 - x = 70 + 75 - y$. So, we get $y = 45 + x$.

Since we want to have the smallest value of y, we let x be 0 and hence get y =45. Therefore, (E) is the correct answer.

解 析

根据题目，全集 U 满足 $n(U) = 100$。设 A 为学习物理的学生的集合，B 为学习化学的学生的集合，x 为不学习这两门课程的学生人数，y 为同时学习这两门课程的学

生人数。利用两个事件的并集公式可得

$$n(A \cup B) = n(A) + n(B) - n(A \cap B)$$

所以 $100 - x = 70 + 75 - y$。因此，我们得到 $y = 45 + x$。

为了使 y 最小，令 $x = 0$，因此得到 $y = 45$。所以 (E) 是正确答案。

Example 8

A survey was done at an outdoor music festival on music genre preference. 78% of the people surveyed like Jazz Music, 80% like Classical Music, 84% like Country Music and 88% like Pop Music. At least how many percent of people surveyed like all four types of music?

例题 8

在一个室外音乐节做一个有关音乐喜好的调查。在受访者中，78% 的人喜欢爵士乐，80% 的人喜欢古典音乐，84% 的人喜欢乡村音乐，88% 的人喜欢流行音乐。问至少有多少受访者同时喜欢 4 种音乐？

(A) 30　　(B) 66　　(C) 83　　(D) 12　　(E) 82.5

Solution

The least percent of people in favor of all four types of music is equal to the difference of 1 and the greatest percent of people who do not like at least one of the four types of music.

If we let the number of people surveyed be 100, we can calculate:

the number of people who do not like Jazz Music:100 – 78=22.

The number of people who do not like Classical Music: 100 – 80=20.

The number of people who do not like Country Music: 100 – 84=16.

And the number of people who do not like Pop Music:100 – 88=12.

Thus, the largest number of people who do not like at least one of the four types of music is: $22+20+16+12=70$.

Therefore, at least $100-70=30$ people, which is equivalent to 30 percent of people, like all four types of music. (A) is the correct answer.

解 析

同时喜欢 4 种音乐的观众的百分比最小值等于 1 减去至少不喜欢某一种音乐的观众的百分比最大值。

设被采访的人数为 100，可得：

不喜欢爵士乐的人数为 $100-78=22$。

不喜欢古典音乐的人数为 $100-80=20$。

不喜欢乡村音乐的人数为 $100-84=16$。

不喜欢流行音乐的人数为 $100-88=12$。

因此，不喜欢至少一种音乐的人数最大为 $22+20+16+12=70$。

因此，至少有 $100-70=30$ 人同时喜欢 4 种音乐，相当于 30% 的人同时喜欢 4 种音乐。(A) 是正确答案。

PRACTICE 练习

Problem 1

A survey was done at the basketball court about which teams they support during the NBA Finals this year. In total, 95 people provided responses. 28 support neither Team A nor Team B because they only support their home teams, while 20 support both Team A and Team B. If 25 viewers support Team A only, how many surveyors support Team B only?

问题 1

一项问卷调查是关于今年支持哪支球队进入 NBA 总决赛，共有 95 人参与了问卷调查。其中 28 人既不支持球队 A 也不支持球队 B，因为他们只支持自己的主队，20 人同时支持球队 A 和球队 B。如果有 25 人只支持球队 A，那么有多少人只支持球队 B？

(A) 3　　(B) 10　　(C) 15　　(D) 22　　(E) 35

Problem 2

Of 40 people, 28 wear gloves and 16 wear scarves during winter. It is also known that 10 wear both gloves and scarves. How many among the 40 wear neither gloves nor scarves?

问题 2

在 40 个人中，有 28 个人冬天戴手套，16 个人冬天带围巾。已知其中有 10 个人既戴手套又戴围巾。那么在这 40 个人中有多少人既不戴手套也不戴围巾？

(A) 10　　(B) 8　　(C) 2　　(D) 6　　(E) 4

Problem 3

How many integers between 1 and 500 (includes 1 and 500) inclusive are not divisible by neither 2, nor 3, nor 5?

问题 3

在 1 到 500 之间（含 1 和 500）的整数中，有多少个整数不能被 2、3 或 5 整除？

(A) 120　　(B) 134　　(C) 143　　(D) 167　　(E) 192

Problem 4

Of 20 boxes, 15 contain pencils, 14 contain pens, and 12 contain both pens and pencils. How many boxes contain neither pens nor pencils?

问题 4

在 20 个盒子中，有 15 个盒子装有铅笔，14 个盒子装有钢笔，12 个盒子同时装有铅笔和钢笔。那么有多少个盒子既没装钢笔也没装铅笔？

(A) 3　　(B) 7　　(C) 6　　(D) 8　　(E) 5

Problem 5

In a class of 50 students, 28 take Biology, 21 take English Literature, and 5 students take both classes. How many students take neither class?

问题 5

在 50 名学生中，有 28 名学生学习生物，21 名学生学习英国文学，5 名学生同时学习这两门课程。有多少名学生两门课都不学习？

(A) 6　　(B) 11　　(C) 45　　(D) 7　　(E) 23

Problem 6

In a survey of 120 students in Elite High School, 84 liked Computer Science, 73 liked Math, and 23 liked Computer Science but not Math. How many students disliked both subjects?

问题 6

在一次对 120 名英才高中学生的调查中，84 人喜欢计算机，73 人喜欢数学，23 人喜欢计算机但不喜欢数学。有多少学生两门学科都不喜欢？

(A) 26 (B) 31 (C) 40 (D) 24 (E) 25

Problem 7

How many numbers between 1 and 1000 are integer multiples of 4 or 6 but not 12?

问题 7

在 1 到 1000 之间，有多少个数是 4 或 6 的倍数但不是 12 的倍数？

(A) 250 (B) 284 (C) 333 (D) 185 (E) 315

Problem 8

71 people subscribed to three different video channels A, B, and C. 40 people subscribed to A, 56 subscribed to B, and 34 subscribed to C. 22 subscribed to A and B. 18 subscribed to B and C. 22 subscribed to A and C. How many people subscribed to all three?

问题 8

71 人订阅了 3 种杂志 A、B 和 C。其中，40 人订阅了 A，56 人订阅了 B，34 人订阅了 C，22 人订阅了 A 和 B，18 人订阅了 B 和 C，22 人订阅了 A 和 C。请问有多少人订阅了全部三种杂志？

(A) 3 (B) 5 (C) 8 (D) 10 (E) 12

Problem 9

Alex, Bob and Charlie are watering 100 flowerpots. Alex watered 86 pots, Bob watered 59 pots, and Charlie watered 75 pots. At least how many flowerpots have been watered three times?

问题 9

亚历克斯、鲍勃和查理正在浇灌 100 盆花。亚历克斯浇了 86 盆，鲍勃浇了 59 盆，查理浇了 75 盆。至少有多少盆花被浇了 3 次？

(A) 14 (B) 16 (C) 20 (D) 31 (E) 40

Problem 10

There are 52 students in Class A. 30 of them like swimming, 35 like playing basketballs, 42 like playing badminton. At least how many students like all three sports?

问题 10

A 班里有 52 名学生。其中，30 名学生喜欢游泳，35 名学生喜欢打篮球，42 名学生喜欢打羽毛球。至少有多少名学生 3 项运动都喜欢？

(A) 7　　(B) 8　　(C) 6　　(D) 3　　(E) 30

Problem 11

Of the 400 students at Northlight High school, 117 take Math, 109 take Economics, and 114 take History. Furthermore, 70 take both Math and Economics, 34 take both Math and History, and 29 take both Economics and History. Finally, 164 students take none of these courses. How many students take all three courses?

问题 11

在北极光高中的 400 名学生中，117 人学习数学，109 人学习经济，114 人学习历史。此外，70 人同时学习数学和经济，34 人同时学习数学和历史，29 人同时学习经济和历史。最后，还有 164 名学生这些课程都没有学。请问有多少学生这 3 门课程都学？

(A) 28　　(B) 29　　(C) 34　　(D) 35　　(E) 63

Problem 12

There are 100 7th graders in the school. 58 like Chinese food, 38 like Italian food, and 52 like American food. 6 students like Italian and Chinese food only (not American food), 4 students like Italian and American food only (not Chinese food), and 12 students like all three types of food. Each student likes at least one type of food. How many students only like Chinese food?

问题 12

学校里有 100 名七年级的学生。58 人喜欢中餐，38 人喜欢意大利餐，52 人喜欢美国餐。有 6 人只喜欢意大利餐和中餐（不喜欢美国餐），4 人只喜欢意大利餐和美国餐（不喜欢中餐），12 人这三种国家的食物都喜欢。每人至少喜欢一个国家的食物。问有多少人只喜欢中餐？

(A) 18　　(B) 19　　(C) 24　　(D) 25　　(E) 26

Problem 13

In a company of 100 employees, 75 of them enjoy participating in team-building exercises, 80 enjoy attending strategy workshops, 92 enjoy social events, and 85 enjoy fitness sessions. What is the minimum number of employees who enjoy all four activities?

问题 13

一家公司有 100 名员工，其中 75 人喜欢参加团建活动，80 人喜欢参加策略工坊，92 人喜欢参加社交活动，85 人喜欢参加健身活动。最少有多少员工这 4 项活动都喜欢?

(A) 55 (B) 77 (C) 35 (D) 32 (E) 19

Problem 14

Set A consists of the positive multiples of 4 that are less than 100, and set B consists of the positive multiples of 6 that are less than 100. How many numbers do sets A and B have in common?

问题 14

集合 A 由小于 100 的 4 的倍数组成，集合 B 由小于 100 的 6 的倍数组成。集合 A 和集合 B 有多少个公共的数?

(A) 0 (B) 1 (C) 2 (D) 4 (E) 8

Problem 15

Set A consists of m integers, and the difference between the greatest integer in A and the least integer in A is 500. A new set of m integers set B, is formed by multiplying each integer in A by 3 and then adding 10 to each of the products. What is the difference between the greatest integer in B and the least integer in B?

问题 15

集合 A 由 m 个整数组成，A 中最大的整数与最小的整数相差 500。通过将 A 中的每个整数乘以 3 然后加上 10，形成一个新的 m 个整数的集合 B。集合 B 中最大的整数与最小的整数的差是多少?

(A) 800 (B) 1000 (C) 2500 (D) 1500 (E) 2400

Problem 16

In a company of 500 employees, each employee has experience in at least one of three key project areas. Here are the details:

260 employees have experience in project management.

260 have experience in data analysis.

75 have experience in both project management and software development.

115 have experience in both software development and data analysis.

130 have experience in both project management and data analysis.

45 have experience in all three areas: project management, data analysis, and software development.

How many employees in this company have experience only in software development?

问题 16

在一家有 500 名员工的公司中，每位员工至少在三个关键领域中的一个有经验。具体情况如下：

260 名员工拥有项目管理经验。

260 名员工拥有数据分析经验。

75 名员工同时拥有项目管理和软件开发经验。

115 名员工同时拥有软件开发和数据分析经验。

130 名员工同时拥有项目管理和数据分析经验。

45 名员工在项目管理、数据分析和软件开发三个领域均有经验。

请问在这家公司中，只有软件开发经验的员工有多少人？

(A) 190 (B) 110 (C) 255 (D) 390 (E) 45

Problem 17

In a town of 350 adults, every adult owns a bike, motorcycle, or both. If 333 adults own bikes and 48 adults own motorcycles, how many of the bike owners do not own a motorcycle?

问题 17

在一个有 350 名成年人的小镇上，每个成年人都有自行车、摩托车或两者都有。如果 333 人有自行车，48 人有摩托车，请问有多少人有自行车但没有摩托车？

(A) 13 (B) 25 (C) 302 (D) 306 (E) 350

Problem 18

Each of the 250 freshmen of Phoenix College has a dog or a cat or both a dog and a cat.

126 students have a dog, and 170 students have a cat. How many students have both a dog and a cat?

问题 18

在菲尼克斯学院一年级 250 名学生中，每个人都有一只狗或一只猫或同时有狗和猫。其中 126 名学生有狗，170 名学生有猫。请问多少名学生同时有狗和猫?

(A) 44 (B) 46 (C) 90 (D) 80 (E) 124

Chapter 8 Triangles and Similarity

第 8 章 三角形与相似性

8.1 Basic Knowledge 基础内容

8.1.1 Abbreviations 缩写

英文	中文	字母
Perimeter	周长	P
Area	面积	A 或 S
Length	长度	l
Width	宽度	w
Height	高度	h
Circumference	圆的周长	C
Raddius	半径	r
Side lengths of a polygon	多边形的边长	$a, b, c, \ldots$

8.1.2 ANGLES: Measure – Sum Principles 角度：度量 – 和定理

Theorem 1 The sum of the measures of the angles of a triangle equals the measure of a straight angle, or 180°.

$$\angle A + \angle B + \angle C = 180°$$

定理 1 三角形内角之和等于一个平角的角度，即 180°。

$$\angle A + \angle B + \angle C = 180°$$

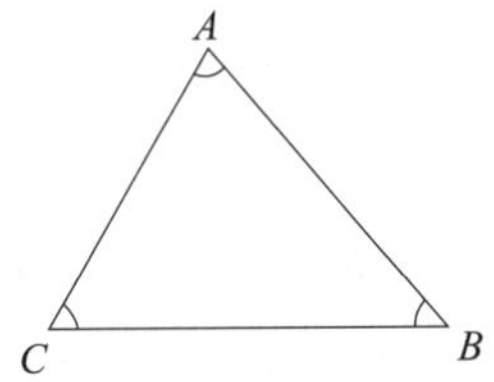

Proof:

We draw the line m parallel to BC, the base of the triangle.

So $\angle C=\angle x, \angle B=\angle y$.

Since $\angle A+\angle x+\angle y=180°$, then $\angle A+\angle B+\angle C=180°$.

证明:

我们绘制一条与三角形底边 BC 平行的直线 m。

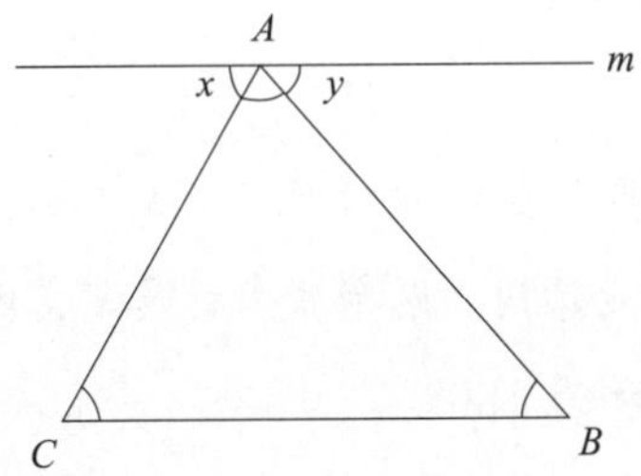

因此 $\angle C=\angle x, \angle B=\angle y$。

由于 $\angle A+\angle x+\angle y=180°$，可得 $\angle A+\angle B+\angle C=180°$。

Example 1

Two angles of an isosceles triangle measure $80°$ and $x°$. What is the sum of the three possible values of x?

例题 1

一个等腰三角形的两个角度分别为 $80°$ 和 $x°$。求 x 的 3 个可能的值之和为多少?

(A) 95　(B) 125　(C) 140　(D) 150　(E) 180

Solution

For isosceles triangle, base angles are identical.

There are 3 cases for $x°$:

Case 1: $x°$ is a base angle with the 80° as the other base angle. In this case, $x = 80°$, since base angles are identical.

Case 2: $x°$ is a base angle with 80° as the vertex angle. The 2 base angles are both $x°$, so we can use the equation $2x + 80 = 180°$, and get $x = 50°$, because the sum of the interior angles of a triangle is 180°.

Case 3: $x°$ is the vertex angle with 80° as a base angle. Both base angles are 80°, since base angles are identical. Thus, we can use the equation $x° + 160° = 180°$ and get $x = 20°$.

Collecting all the cases, we get 80° + 50° + 20° = 150°, so the answer is (D).

解 析

根据等腰三角形性质，底角相等。

$x°$ 有 3 种情况进行讨论：

情况 1：$x°$ 和 80° 都是底角。在这种情况下，$x = 80$，因为底角相等。

情况 2：$x°$ 是底角，而 80° 是顶角。两个底角都是 $x°$，因此可以使用等式 $2x+80 = 180$，得到 $x = 50$，因为三角形内角和为 180°。

情况 3：$x°$ 是顶角而两个底角都是 80°。两个底角都是 80°，因为底角相同。因此，可以使用等式 $x + 160 = 180$，得出 $x = 20$。

将 3 种情况的值相加，得到 80°+ 50°+ 20° = 150°，因此答案是 (D)。

Theorem 2 The measure of each exterior angle of a triangle equals the sum of the measures of its two remote nonadjacent interior angles.

定理 2 三角形的外角等于其两个不相邻的内角之和。

$$z = a + b,\ x = b + c,\ y = c + a$$

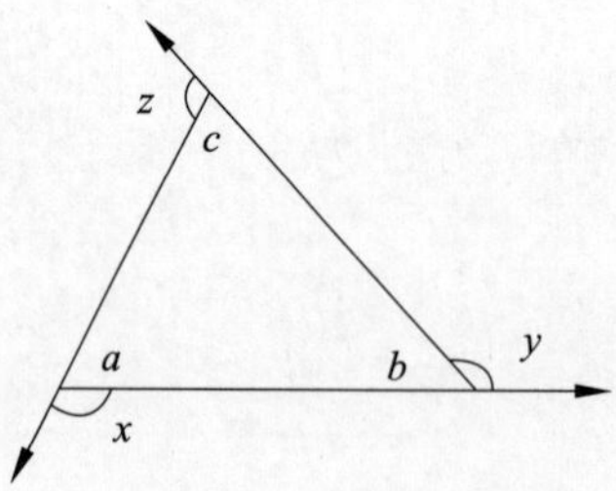

Proof:

$$b + y = 180° \quad ①$$

$$a + b + c = 180° \quad ②$$

①－②:

$$y-(c+a)=0 \Rightarrow y=c+a$$

Similarly, we can prove $x=b+c$ and $z=a+b$.

证明:

$$b+y=180° \quad ①$$

$$a+b+c=180° \quad ②$$

①－②得

$$y-(c+a)=0 \Rightarrow y=c+a$$

同理可证明 $x=b+c$ 和 $z=a+b$。

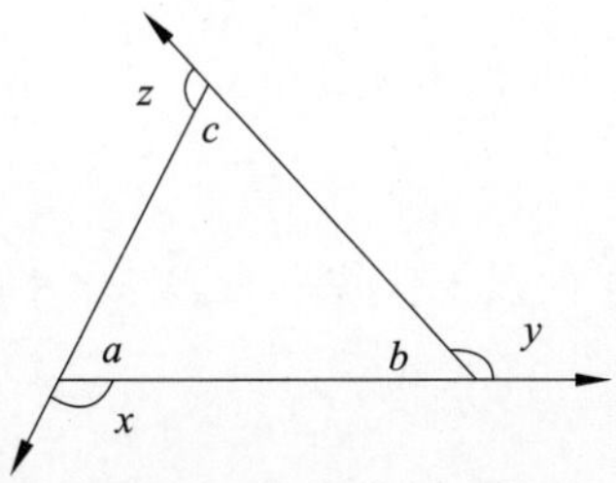

Theorem 3 The sum of the measures of the exterior angles of a triangle equals 360°.

$$x+y+z=360°$$

定理 3 三角形的外角之和为 360°。

$$x+y+z=360°$$

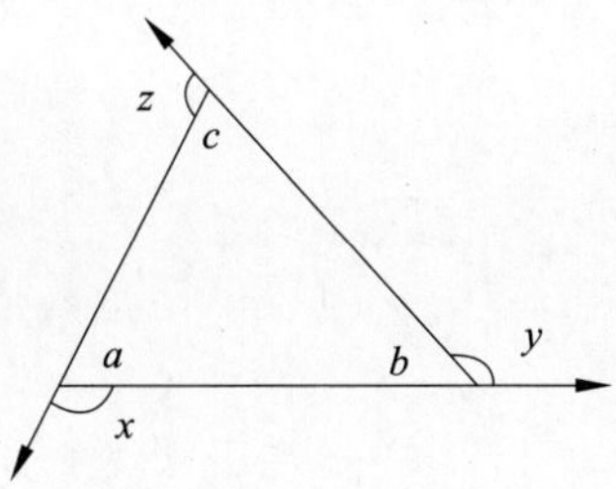

Proof:

$$x=b+c \quad ①$$

$$y=a+c \quad ②$$

$$z=a+b \quad ③$$

$$a+b+c=180° \quad ④$$

① + ② +③:

$$x+y+z=2(a+b+c) \quad ⑤$$

Substituting ④ into ⑤:

$$x + y + z = 360°$$

证明:

$$x = b + c \quad ①$$

$$y = a + c \quad ②$$

$$z = a + b \quad ③$$

$$a + b + c = 180° \quad ④$$

① + ② +③得

$$x + y + z = 2(a + b + c) \quad ⑤$$

将④ 代入 ⑤得

$$x + y + z = 360°$$

Theorem 4 The four angles in the figure below have the following relationship: $\angle D = \angle A + \angle B + \angle C$.

定理 4 下图中的四个角存在以下关系：$\angle D = \angle A + \angle B + \angle C$。

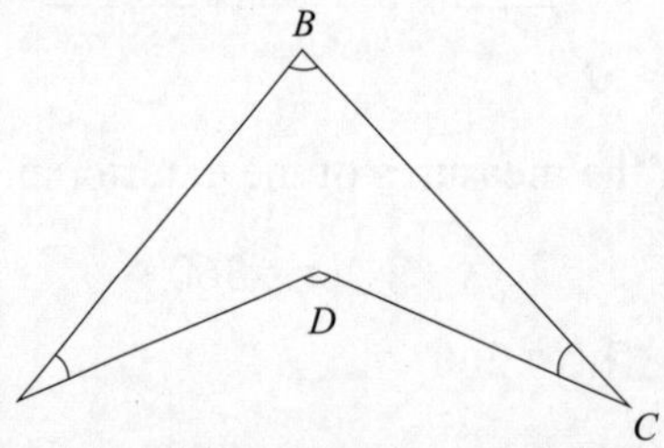

Proof:

Extend AD to meet BC at E.

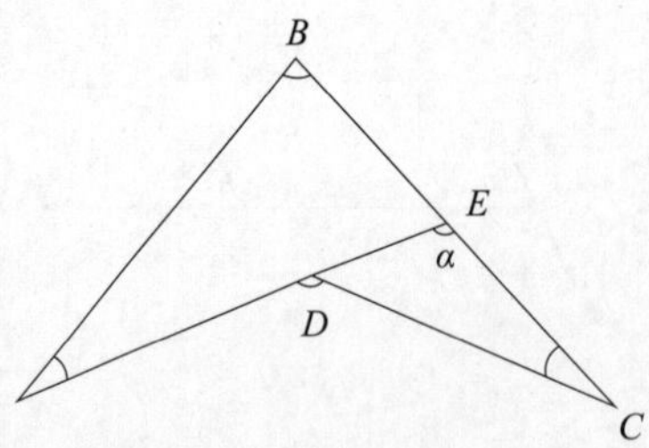

By Theorem 2, $\angle D = \alpha + \angle C$. ①

By Theorem 2, $\alpha = \angle A + \angle B$. ②

Substituting ② into ① : $\angle D = \angle A + \angle B + \angle C$.

证明:

延长 AD 使其交 BC 于 E 点。

根据定理 2，$\angle D = \alpha + \angle C$。 ①

根据定理 2， $\alpha = \angle A + \angle B$ 。 ②

将②代入①得 $\angle D = \angle A + \angle B + \angle C$ 。

8.1.3 Basic Formulas for Perimeters and Areas of Triangle 三角形周长和面积的基本公式

Perimeter of a triangle: $P = a + b + c$.

Area of a triangle: $A = \frac{1}{2}bh_b$, where h_b is the height on the side b.

For an equilateral triangle (three sides have the same length, a), the area is:

$$A = \frac{\sqrt{3}}{4}a^2$$

三角形的周长：$P = a + b + c$。

三角形的面积：$A = \frac{1}{2} \times b \times h_b$，其中高为 h_b，代表的是垂直于边 b 的高。

对于等边三角形（三边长相等，为 a）的面积为

$$A = \frac{\sqrt{3}}{4}a^2$$

8.1.4 The Relationship Between the Three Sides of a Triangle (Triangle Inequality Theorem) 三角形三条边的关系（三角不等式）

(1) In a triangle, the sum of the lengths of any two sides is larger than the length of the third side.

（1）在三角形中，任意两边长度之和必须大于第三边的长度。

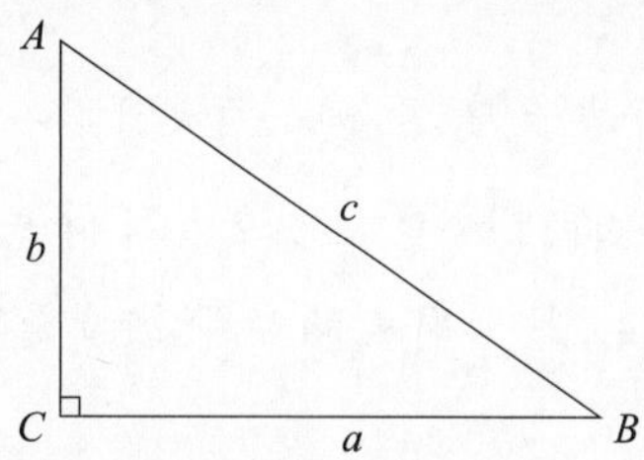

$$a + b > c$$

$$a + c > b$$

$$b + c > a$$

(2) In a triangle, the subtraction between any two sides must be less than the length of the third side.

(2)在三角形中，任意两条边的差值必须小于第三边的长度。

$$|a-b|<c$$
$$|a-c|<b$$
$$|b-c|<a$$

Example 2

What is the smallest whole number larger than the perimeter of any triangle with a side of length 4 and a side of length 20?

例题 2

已知一个三角形两条边长分别为 4 和 20，大于这个三角形最大周长的最小整数是多少？

(A) 24　　(B) 29　　(C) 43　　(D) 48　　(E) 57

Solution

We know from the triangle inequality that the sum of the lengths of any two sides is larger than the length of the third side. Therefore, we can conclude that $s < 4+20 = 24$. Adding $4 + 20$ to both sides of the inequality, we get $s + 4 + 20 < 24 + 24 = 48$. Since $s + 4 + 20$ is the perimeter of our triangle, the smallest whole number larger than the perimeter is 48. The answer is (D).

解　析

根据三角不等式，任意两边长度之和必须大于第三边的长度。因此，我们可以得出 $s<5+19=24$。将 $4+20$ 加到不等式两边，可得 $s+4+20<24+24=48$。由于 $s+4+20$ 是三角形的周长，所以大于周长的最小整数是 48。答案是 (D)。

Example 3

The coordinates of $\triangle ABC$ are $A(5,7)$, $B(11,7)$, and $C(3,y)$, with $y>7$. The area of $\triangle ABC$ is 18. What is the value of y?

例题 3

$\triangle ABC$ 的顶点坐标分别为 $A(5,\ 7)$，$B(11,\ 7)$ 和 $C(3,\ y)$，且 $y>7$。已知 $\triangle ABC$ 的面积为 18。求 y 的值？

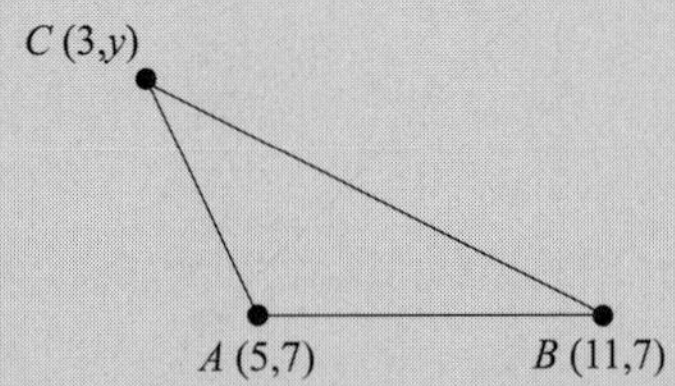

(A) 6 (B) 8 (C) 9 (D) 10 (E) 13

Solution

By reading the coordinates, we know the triangle has base 11 − 5=6. So, its height h satisfies $\frac{1}{2}\times AB\times h=\frac{1}{2}\times 6\times h=18$. This means that $h=6$. Since $y-h=7$, we get that $h=7+6=13$. The answer is (E).

解 析

根据顶点的坐标，可得三角形的底边长度为 11 − 5=6。因此，高 h 满足 $\frac{1}{2}\times AB\times h=\frac{1}{2}\times 6\times h=18$。解得 $h=6$。由于 $y-h=7$，所以 $h=7+6=13$。因此答案是 (E)。

8.2 Pythagorean Theorem 勾股定理

(For right triangles only): $a^2+b^2=c^2$.

(a and b are two legs. c is the hypotenuse.)

（仅适用于直角三角形）：$a^2+b^2=c^2$。

（a 和 b 是两条直角边，c 是斜边。）

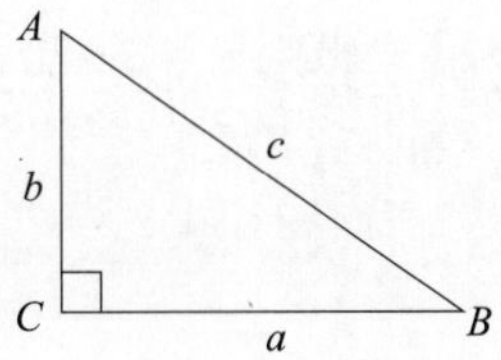

Proof:

In the right triangle ABC, draw $CD \perp AB$.

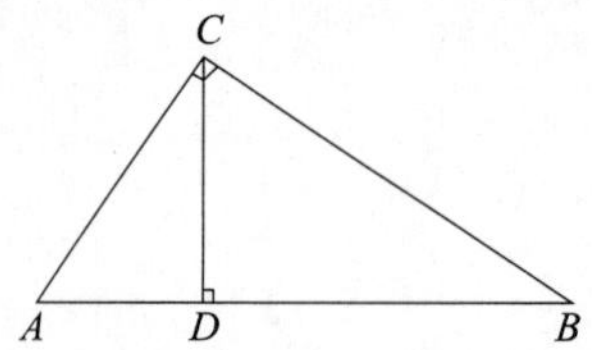

Since $\triangle ABC$, $\triangle ACD$ and $\triangle CBD$ are similar, the edges satisfy

$$\frac{AC}{AB}=\frac{AD}{AC}\Rightarrow AC^2=AB\times AD \quad ①$$

$$\frac{BC}{AB}=\frac{BD}{BC}\Rightarrow BC^2=AB\times BD \quad ②$$

①+②:

$$AC^2+BC^2=AB\times AD+AB\times BD=AB(AD+BD)=AB\times AB=AB^2$$

证明:

在直角三角形 ABC 中，作 $CD\perp AB$。

因为 $\triangle ABC$， $\triangle ACD$ 和 $\triangle CBD$ 相似，所以边的关系满足

$$\frac{AC}{AB}=\frac{AD}{AC}\Rightarrow AC^2=AB\times AD \quad ①$$

$$\frac{BC}{AB}=\frac{BD}{BC}\Rightarrow BC^2=AB\times BD \quad ②$$

①+②:

$$AC^2+BC^2=AB\times AD+AB\times BD=AB(AD+BD)=AB\times AB=AB^2$$

Example 4

The base of isosceles $\triangle ABC$ is 30 and its area is 300. What is the length of one of the congruent sides?

例题 4

等腰 $\triangle ABC$ 的底边长为 30，面积为 300。求腰的长度。

(A) 20 (B) 25 (C) 27.5 (D) 30 (E) 35

Solution

The area of a triangle is calculated by $\frac{1}{2}\times\text{base}\times\text{height}$. Given that the base equals to 30 and the area equals to 300, using $\frac{1}{2}\times 30\times\text{height}=300$, we get the height is 20. In this isosceles triangle, the height bisects the base, so by the Pythagorean Theorem, $a^2+b^2=c^2$, we can solve for the hypotenuse c from equation $20^2+15^2=c^2$, and get $c = 25$, which is the length of the congruent sides. The answer is (B).

解 析

三角形的面积计算公式为 $\frac{1}{2}\times$底$\times$高。已知底边为 30，面积为 300，使用公式 $\frac{1}{2}\times 30\times$ 高 =300，可得高为 20。在这个等腰三角形中，高平分底边，根据勾股定理 $a^2+b^2=c^2$，可以通过方程 $20^2+15^2=c^2$，得出 $c = 25$，即腰的长度。因此答案是 (B)。

8.3 Some Theorems Extending Pythagprean Theorem 勾股定理的延伸定理

Theorem 1 When we draw similar figures on the two legs and the hypotenuse of a right triangle, the following formula is true:

$$S_1=S_2+S_3$$

定理 1 在直角三角形的两条直角边和斜边上绘制相似图形时，以下公式成立：

$$S_1=S_2+S_3$$

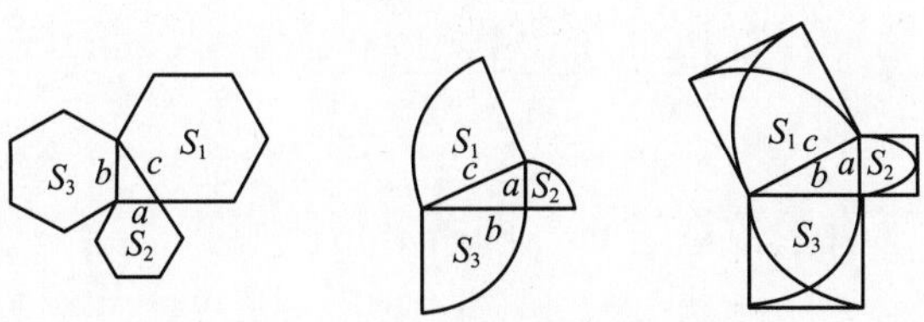

Proof:

$$\frac{S_2}{S_1}=\left(\frac{a}{c}\right)^2$$

$$\frac{S_3}{S_1}=\left(\frac{b}{c}\right)^2$$

$$S_2+S_3=\frac{a^2+b^2}{c^2}S_1=S_1$$

证明:

$$\frac{S_2}{S_1}=\left(\frac{a}{c}\right)^2$$

$$\frac{S_3}{S_1}=\left(\frac{b}{c}\right)^2$$

$$S_2+S_3=\frac{a^2+b^2}{c^2}S_1=S_1$$

Example 5

A hexagram is composed of a regular hexagon of side length 2 and its 6 equilateral triangular extensions, as shown in the diagram. What is the ratio of the area of the extensions to the area of the original hexagon?

例题 5

如图所示，一个六芒星由一个边长为 2 的正六边形及其 6 个等边三角形延展部分组成。请问延展部分的面积与原正六边形面积的比是多少?

(A) 1 : 1　　(B) 6 : 5　　(C) 3 : 2　　(D) 2 : 1　　(E) 3 : 1

Solution

By connecting the three pairs of opposite vertices, the regular hexagon is cut into 6 equilateral triangles with side length 2. Therefore, the ratio of area of the extensions to the area of the original hexagon is equivalent to $\frac{6\times \text{area of small equilateral triangle}}{6\times \text{area of small equilateral triangle}}=1$, so the answer is (A).

解 析

通过连接三组相对的顶点，可以将正六边形切割成 6 个边长为 2 的等边三角形。因此，扩展部分的面积与原六边形的面积之比为$\frac{6\times 小等边三角形的面积}{6\times 小等边三角形的面积}=1$，故答案是 (A)。

Theorem 2 In a 45°– 45°– 90°right triangle, or right isosceles triangle, the length of the hypotenuse is $\sqrt{2}$ times of the length of each leg.

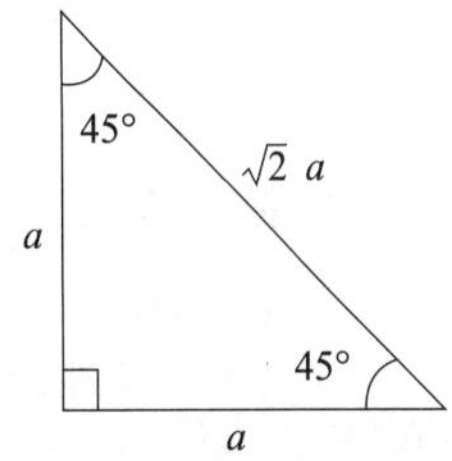

定理 2 在 45°– 45°– 90° 的直角三角形，或等腰直角三角形中，斜边的长度是直角边长度的 $\sqrt{2}$ 倍。

Example 6

$\triangle ABC$ is isosceles with $AC= BC$. Angle A measures 45 degrees. Segment CD is the perpendicular bisector of segment AB. If segment AD measures 5, how long is segment AC?

例题 6

$\triangle ABC$ 是等腰三角形，$AC = BC$。$\angle A$ 为 45°。线段 CD 是线段 AB 的垂直平分线。如果线段 AD 的长度为 5，求线段 AC 的长度？

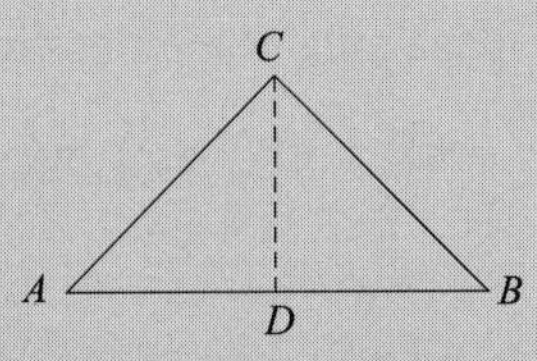

(A) 5 (B) 10 (C) $5\sqrt{2}$ (D) $5\sqrt{3}$ (E) $6\sqrt{2}$

Solution

Since ABC is isosceles with $AC = BC$, we know $\angle A = \angle B = 45°$. As such, $\angle ACB = 90°$. Since CD is the perpendicular bisector of segment AB, we have $CD \perp AD$, and $\angle CDA = \angle CDB = 90°$. Therefore, $\angle DCA = \angle A = 45°$ and $AD = DC$, get that ACD is an isosceles right triangle.

By Theorem 2, $AC = \sqrt{2} \times 5 = 5\sqrt{2}$ due to $AD = 5$. The answer is (C).

解　析

已知$\triangle ABC$是等腰三角形，$AC = BC$，我们知道$\angle A = \angle B = 45°$。因此$\angle ACB = 90°$。又因为 CD 是线段 AB 的垂直平分线，$CD \perp AD$，于是$\angle CDA = \angle CDB = 90°$。因此$\angle DCA = \angle A = 45°$，且$AD = DC$，得到$\triangle ACD$是等腰直角三角形。

根据定理 2，因为 $AD = 5$，所以 $AC = \sqrt{2} \times 5 = 5\sqrt{2}$。答案是 (C)。

Theorem 3　For a right triangle, if $\angle A = 30°$, then $BC = \frac{1}{2}AB$.

定理 3　直角三角形中，如果 $\angle A = 30°$，那么 $BC = \frac{1}{2}AB$。

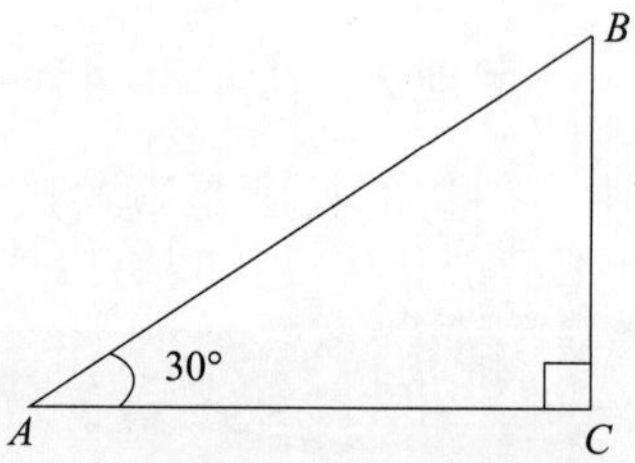

Example 7

If $\angle B = 60°$ in right triangle ABC, and $BC = 2\sqrt{3}$, find AC.

例题 7

在直角$\triangle ABC$中，如果$\angle B = 60°$，且$BC = 2\sqrt{3}$，求 AC 的长度。

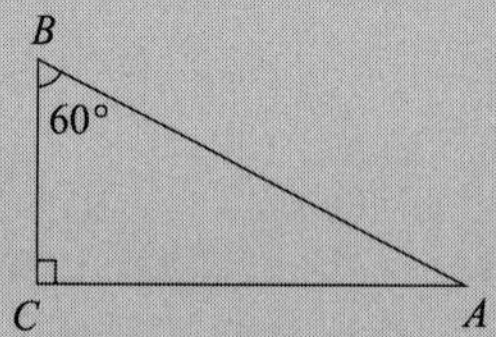

(A) $4\sqrt{2}$　　(B) $4\sqrt{3}$　　(C) 6　　(D) $6\sqrt{2}$　　(E) 18

Solution

Since $\angle B$=60°, $\angle C$=90°, we know that $\angle A$=30°. By Theorem 3, $AB=2BC=4\sqrt{3}$. By the Pythagorean Theorem, $AC^2+BC^2=AB^2$, therefore $AC^2=AB^2-BC^2=(4\sqrt{3})^2-(2\sqrt{3})^2=36$ and we can get $AC=6$. The answer is (C).

解 析

由于$\angle B$=60°，且$\angle C$=90°，因此$\angle A$=30°。根据定理 3，$AB=2BC=4\sqrt{3}$。根据勾股定理，$AC^2+BC^2=AB^2$，因此$AC^2=AB^2-BC^2=(4\sqrt{3})^2-(2\sqrt{3})^2=36$，可得$AC=6$。所以答案是 (C)。

Theorem 4 The length of the median CM to the hypotenuse of a right triangle equals one-half the length of the hypotenuse.

定理 4 在直角三角形中，从点 C 引一条斜边的中线 CM，则长度为斜边的一半。

$$AM = MB = MC$$

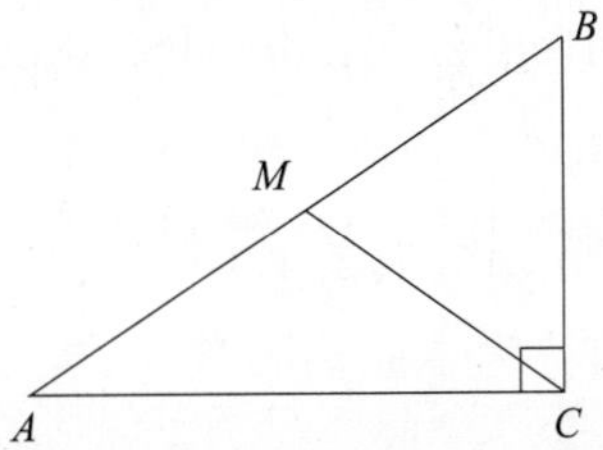

Example 8

If $\triangle QPR$ is a right triangle where M is the midpoint of RQ, and $MP=10$, the length of RQ is?

例题 8

如果$\triangle QPR$是一个直角三角形，其中 M 是线段 RQ 的中点，且 $MP=10$，求 RQ 的长度。

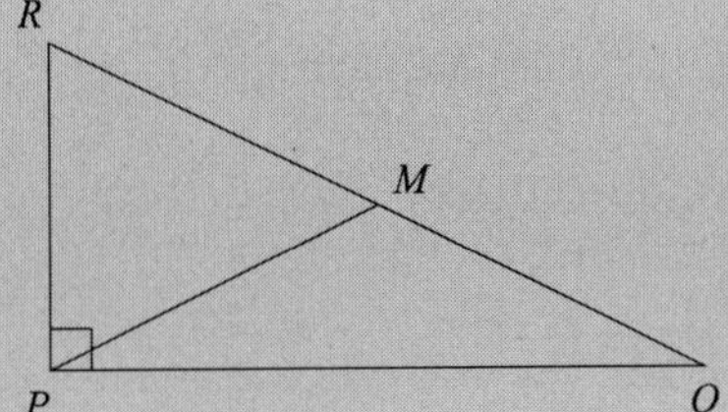

(A) 10 (B) 20 (C) $10\sqrt{2}$ (D) $20\sqrt{2}$ (E) 30

Solution

Since M is the midpoint of RQ, MP is the median to the hypotenuse RQ. By Theorem 4, we know $RM=MQ=MP=10$. Therefore, $RQ = RM + MQ = 2MP=20$. The answer is (B).

解 析

由于 M 是线段 RQ 的中点，MP 是斜边 RQ 的中线。根据定理 4，$RM=MQ=MP=10$。因此 $RQ = RM + MQ = 2MP = 20$。答案是 (B)。

8.4 Pythagorean Triples 勾股数

A Pythagorean triple is an ordered triple (a, b, c) of three positive integers such that $a^2+b^2=c^2$. If a, b and c are relatively prime, then the triple is called primitive.

Here is a table of some primitive Pythagorean triples:

勾股数是一个是由三个正整数组成的有序三元组 (a，b，c)，其中 $a^2+b^2=c^2$。如果 a，b，c 互质，则该三元组称为本原勾股数。

以下是一些本原勾股数：

a	b	c	a	b	c
3	4	5	11	60	61
5	12	13	12	35	37
7	24	25	13	84	85
8	15	17	15	112	113
9	40	41	20	21	29

Example 9

Let A be the area of the triangle with sides of length 30, 30 and 36. Let B be the area of the triangle with sides of length 30, 30 and 48. What is the relationship between A and B?

例题 9

设 A 为边长为 30，30 和 36 的三角形的面积。B 为边长为 30，30 和 48 的三角形的面积。问 A 和 B 之间的关系是什么？

(A) $A=\frac{9}{16}B$ (B) $A=\frac{3}{4}B$ (C) $A=B$ (D) $A=\frac{4}{3}B$ (E) $A=\frac{16}{9}B$

Solution

For the triangle with side of length 30, 30 and 36, we can draw the altitude to the side with length 36. By HL Congruence (in two right triangles the hypotenuse and one leg are equal, then the triangles are congruent), the two triangles formed are congruent. Thus the altitude splits the side with length 36 into two segments with length 18. By the Pythagorean Theorem, we have $18^2+x^2=30^2$, so we get the height $x=24$. For the triangle with side of length 30, 30 and 36, the area is $A=\frac{1}{2}\times 30\times 24=432$.

For the triangle with side of length 30, 30 and 48, we can draw the altitude to the side with length 48. Similarly, by HL Congruence, the two triangles formed are congruent. Thus, the altitude splits the side with length 48 into two segments with length 24. From the previous case, we know that the other side length is 18 as $(18,24,30)$ is a Pythagorean triple. For the triangle with side of length 30, 30 and 48, the area is $B=\frac{1}{2}\times 48\times 18=432$.

Therefore, A=B. The answer is (C).

解 析

对于边长为 30，30 和 36 的三角形，做一条到边长 36 的垂线为三角形的高。根据 HL 定理（在两个直角三角形中，如果斜边和一条直角边相等，则两个三角形全等），两个三角形全等。因此，高将边长为 36 的边分成两个长为 18 的部分。根据勾股定理，$18^2+x^2=30^2$。解得 $x=24$。因此，边长为 30，30 和 36 的三角形的面积为 $A=\frac{1}{2}\times 30\times 24=432$。

对于边长为 30，30 和 48 的三角形，做一条到边长 40 的垂线为三角形的高。同理，根据 HL 定理，两个三角形全等。因此，高将边长为 48 的边分成两个长为 24 的部分。上一种情况，可得另一边的长度为 18，因为 $(18, 24, 30)$ 是一组勾股数。因此，边长为 30，30 和 48 的三角形的面积为 $B=\frac{1}{2}\times 48\times 18=432$。所以 $A=B$。答案是 (C)。

8.5 Similar Triangles 相似三角形

Similar triangles are triangles whose corresponding angles are congruent and whose corresponding sides are in proportion to each other. Similar triangles have the same shape but are not necessarily the same size.

The symbol for "similar" is $\backsim$. The notation $\triangle ABC \backsim \triangle A'B'C'$ is read "$\triangle ABC$ is similar to $\triangle A'B'C'$".

相似三角形是指对应角相等且对应边成比例的三角形。相似三角形具有相同的形状，但不一定大小相同。

"相似"的符号是$\backsim$。这个符号 $\triangle ABC \backsim \triangle A'B'C'$ 应被读为"$\triangle ABC$ 与 $\triangle A'B'C'$ 相似"。

8.5.1 Principles of Similar Triangles 相似三角形的判定

Principle 1 (SSS) Corresponding sides (segments) of similar triangles are in proportion to each other.

If $\triangle ABC \backsim \triangle A_1B_1C_1$, then $\frac{a}{a_1}=\frac{b}{b_1}=\frac{c}{c_1}$.

If $\frac{a}{a_1}=\frac{b}{b_1}=\frac{c}{c_1}$, then $\triangle ABC \backsim \triangle A_1B_1C_1$.

定理 1 （SSS）相似三角形的对应边（线段）成比例。

如果 $\triangle ABC \backsim \triangle A_1B_1C_1$，则 $\frac{a}{a_1}=\frac{b}{b_1}=\frac{c}{c_1}$。

如果 $\frac{a}{a_1}=\frac{b}{b_1}=\frac{c}{c_1}$，则 $\triangle ABC \backsim \triangle A_1B_1C_1$。

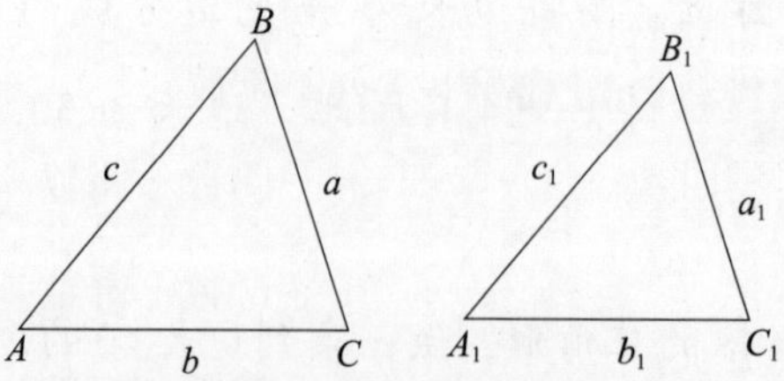

Principles 2 (AA) If two angles of one triangle are congruent respectively to two angles of the other triangle, the two triangles are similar by AA (angle, angle).

Corollary of Principle 2 Two right triangles are similar if they have one congruent acute angle.

定理 2 （AA）如果一个三角形的两个角分别与另一个三角形的两个角相等，则根据 AA（角角）定理，这两个三角形相似。

定理 2 的推论 如果两个直角三角形有一个相同的锐角，则这两个直角三角形相似。

Principles 3 (SAS) If two sides of one triangle are proportional to the corresponding parts of another triangle, and the included angles are congruent, the two triangles are similar by SAS (side, angle, side).

If $\frac{a}{c}=\frac{b}{d}$ and $\alpha=\beta$, then two triangles are similar.

定理 3 （SAS）如果一个三角形的两条边与另一个三角形的对应部分成比例，且夹角相等，则根据 SAS（边角边）原理，这两个三角形是相似的。

如果 $\frac{a}{c}=\frac{b}{d}$， $\alpha=\beta$，则这两个三角形相似。

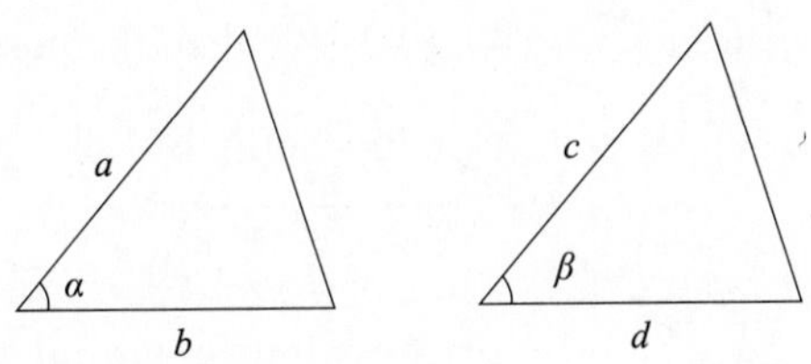

8.5.2 Important Theorems 重要定理

Theorem 1 The ratio of the perimeters of two similar triangles is

$$\frac{P_{\triangle ABC}}{P_{\triangle A_1B_1C_1}}=\frac{a}{a_1}=\frac{b}{b_1}=\frac{c}{c_1}$$

The ratio of the areas of two similar triangles is

$$\frac{S_{\triangle ABC}}{S_{\triangle A_1B_1C_1}}=\left(\frac{a}{a_1}\right)^2=\left(\frac{b}{b_1}\right)^2=\left(\frac{c}{c_1}\right)^2$$

定理 1 两个相似三角形的周长比为

$$\frac{P_{\triangle ABC}}{P_{\triangle A_1B_1C_1}}=\frac{a}{a_1}=\frac{b}{b_1}=\frac{c}{c_1}$$

两个相似三角形的面积比为

$$\frac{S_{\triangle ABC}}{S_{\triangle A_1B_1C_1}}=\left(\frac{a}{a_1}\right)^2=\left(\frac{b}{b_1}\right)^2=\left(\frac{c}{c_1}\right)^2$$

Theorem 2 A line parallel to a side of a triangle cuts off a triangle similar to the given triangle. If $DE//BC$, then $\triangle ABC \backsim \triangle ADE$.

定理 2 一条平行于三角形一边的直线切割三角形并形成与原三角形相似的三角形。如果 $DE//BC$，那么 $\triangle ABC \backsim \triangle ADE$。

Proof:

Since $DE//BC$, $\angle ADE=\angle ABC$ and $\angle AED=\angle ACB$.

By the Principle 2 (AA), $\angle ADE=\angle ABC$, then $\triangle ABC \backsim \triangle ADE$.

If $\triangle ABC \backsim \triangle ADE$, then $\frac{AD}{AB}=\frac{AE}{AC}=\frac{DE}{BC}, \frac{AD}{DB}=\frac{AE}{EC}, \frac{AD}{AE}=\frac{DB}{EC}$.

证明：

因为 $DE//BC$，所以 $\angle ADE=\angle ABC$ 且 $\angle AED=\angle ACB$。

根据定理 2（AA），$\angle ADE=\angle ABC$，所以 $\triangle ABC \backsim \triangle ADE$。

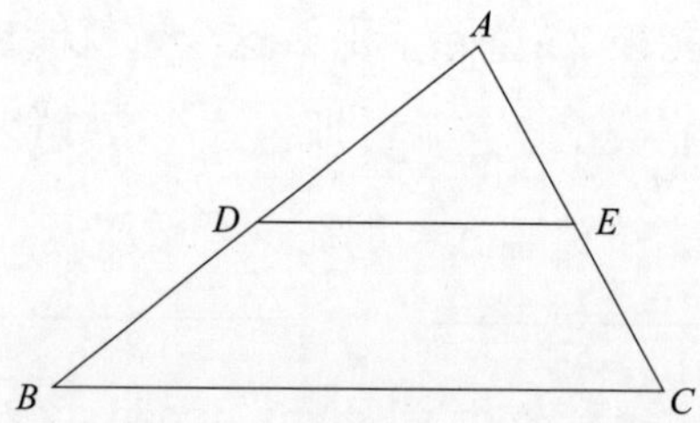

如果 $\angle ADE=\angle ABC$，那么 $\frac{AD}{AB}=\frac{AE}{AC}=\frac{DE}{BC}, \frac{AD}{DB}=\frac{AE}{EC}, \frac{AD}{AE}=\frac{DB}{EC}$。

Theorem 3 In $\triangle ABC$, if D is the midpoint of AB, E is the midpoint of AC, then $DE//BC$ and $DE=\frac{1}{2}BC$.

定理 3 在 $\triangle ABC$ 中，如果 D 是 AB 的中点，E 是 AC 的中点，则 $DE//BC$ 且 $DE=\frac{1}{2}BC$。

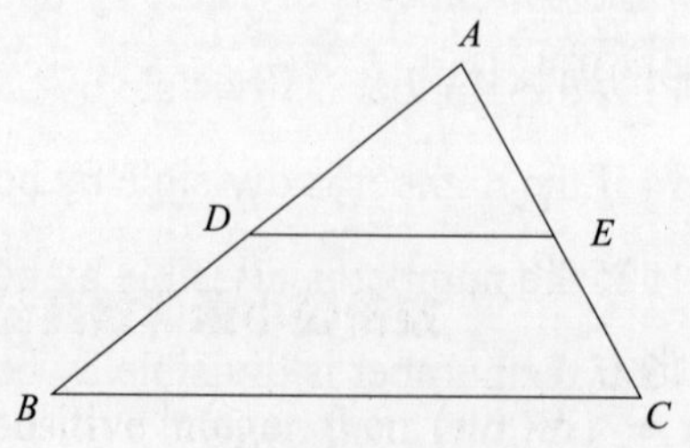

Proof:

We see that $\dfrac{AD}{AB}=\dfrac{1}{2}$, $\dfrac{AE}{AC}=\dfrac{1}{2}$, and $\angle A=\angle A$.

By the Principles 3 (SAS), we know that $\triangle ABC \backsim \triangle ADE$. So $\angle ADE=\angle ABC$ and $\angle AED=\angle ACB$. Thus, $DE//BC$.

$\dfrac{AD}{AB}=\dfrac{DE}{BC}=\dfrac{1}{2}\Rightarrow DE=\dfrac{1}{2}BC$.

证明:

我们知道 $\dfrac{AD}{AB}=\dfrac{1}{2}$，$\dfrac{AE}{AC}=\dfrac{1}{2}$，$\angle A=\angle A$。

根据定理 3（SAS），我们知道 $\triangle ABC \backsim \triangle ADE$。所以 $\angle ADE=\angle ABC$ 且 $\angle AED=\angle ACB$。因此，$DE//BC$。

$\dfrac{AD}{AB}=\dfrac{DE}{BC}=\dfrac{1}{2}\Rightarrow DE=\dfrac{1}{2}BC$。

Theorem 4 In $\triangle ABC$, if D is the midpoint of AB, and $DE//BC$, then E is the midpoint of AC and $DE=\dfrac{1}{2}BC$.

定理 4 在 $\triangle ABC$ 中，如果 D 是 AB 的中点，且 $DE//BC$，那么 E 是 AC 的中点，且 $DE=\dfrac{1}{2}BC$。

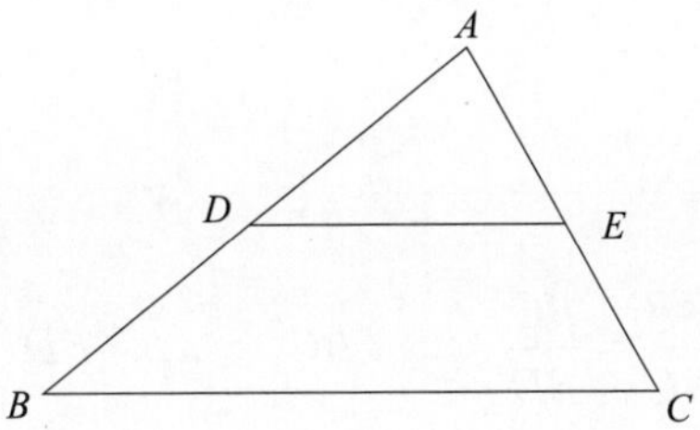

Theorem 5 If $\angle ACB=\angle ADC=90°$, then $\triangle ABC \backsim \triangle ACD \backsim \triangle CBD$.

$$AC^2=AB\times AD \quad ①$$

$$BC^2=AB\times BD \quad ②$$

$$CD^2=AD\times BD \quad ③$$

$$CD\times AB=AC\times BC \quad ④$$

定理 5 如果 $\angle ACB=\angle ADC=90°$，那么 $\triangle ABC \backsim \triangle ACD \backsim \triangle CBD$。

$$AC^2=AB\times AD \quad ①$$

$$BC^2 = AB \times BD \quad ②$$

$$CD^2 = AD \times BD \quad ③$$

$$CD \times AB = AC \times BC \quad ④$$

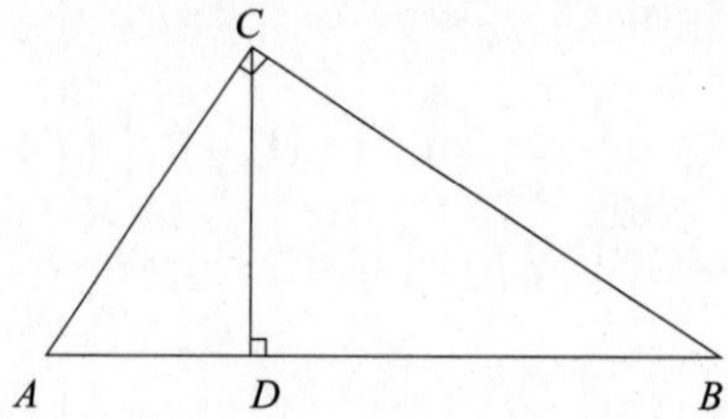

Proof:

(1) We separate two similar triangles as follows:

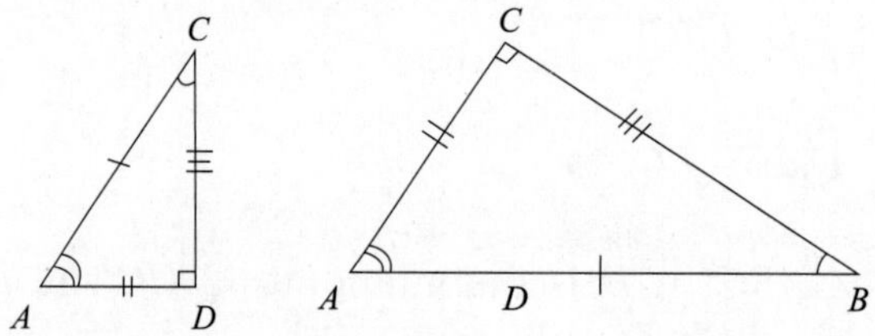

$$\frac{AC}{AB} = \frac{AD}{AC} \Rightarrow AC^2 = AB \times AD \quad ①$$

(2) We separate two similar triangles as follows:

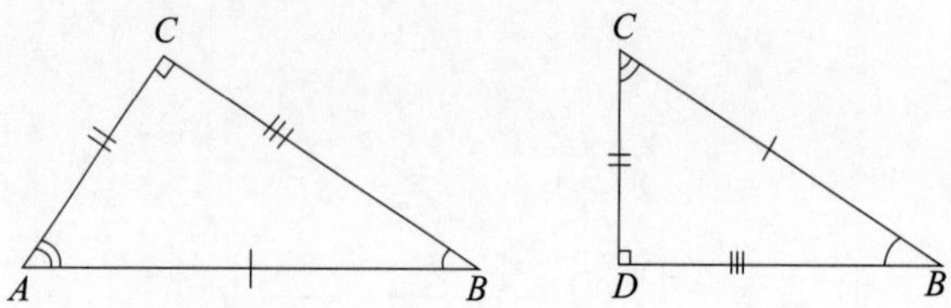

$$\frac{AB}{BC} = \frac{BC}{BD} \Rightarrow BC^2 = AB \times BD \quad ②$$

(3) We separate two similar triangles as follows:

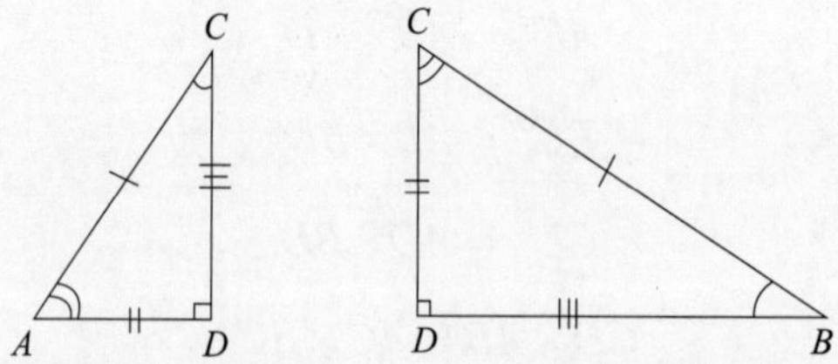

$$\frac{AD}{CD} = \frac{CD}{BD} \Rightarrow CD^2 = AD \times BD \quad ③$$

(4) The area of $\triangle ABC$ is :

$$S_{\triangle ABC}=\frac{1}{2}AC\times BC$$

$S_{\triangle ABC}$ can also be written as $S_{\triangle ABC}=\frac{1}{2}AB\times CD$.

$$\frac{1}{2}AC\times BC=\frac{1}{2}AB\times CD \Rightarrow CD\times AB=AC\times BC \quad ④$$

证明：

（1）我们将两个相似的三角形分开如下：

$$\frac{AC}{AB}=\frac{AD}{AC} \Rightarrow AC^2=AB\times AD \quad ①$$

（2）我们将两个相似的三角形分开如下：

$$\frac{AB}{BC}=\frac{BC}{BD} \Rightarrow BC^2=AB\times BD \quad ②$$

（3）我们将两个相似的三角形分开如下：

$$\frac{AD}{CD}=\frac{CD}{BD} \Rightarrow CD^2=AD\times BD \quad ③$$

（4）$\triangle ABC$ 的面积可以写为

$$S_{\triangle ABC}=\frac{1}{2}AC\times BC$$

$S_{\triangle ABC}$ 也可以写作 $S_{\triangle ABC}=\frac{1}{2}AB\times CD$

$$\frac{1}{2}AC\times BC=\frac{1}{2}AB\times CD \Rightarrow CD\times AB=AC\times BC \quad ④$$

Example 10

In $\triangle ABC$, a point E is on AB with AE=1 and EB=2. Point D is on AC so that $DE//BC$ and point F is on BC so that $EF//AC$. What is the ratio of the area of $CDEF$ to the area of $\triangle EFB$?

例题 10

在 $\triangle ABC$ 中，点 E 位于 AB 上，AE =1 和 EB =2。点 D 位于 AC 上使得 $DE//BC$，点 F 位于 BC 上使得 $EF//AC$。求四边形 $CDEF$ 的面积和 $\triangle EFB$ 的面积的比？

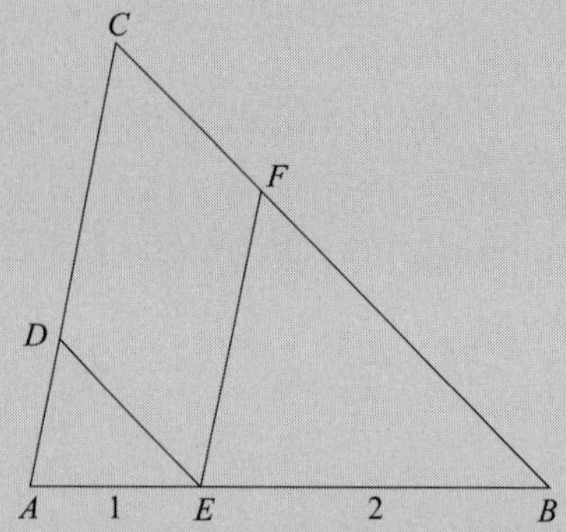

(A) $\frac{4}{5}$ (B) 1 (C) $\frac{5}{4}$ (D) $\frac{2}{3}$ (E) $\frac{4}{9}$

Solution

Since $EF//AC$, $\angle DAE=\angle FEB$. Similarly, since $DE//BC$, $\angle DEA=\angle FBE$.

Therefore, we have $\triangle ADE \backsim \triangle EFB$. We can obtain

$$\frac{\text{area of }\triangle ADE}{\text{area of }\triangle EFB}=\left(\frac{AE}{EB}\right)^2=\left(\frac{1}{2}\right)^2=\left(\frac{1}{4}\right)^2$$

and therefore area of $\triangle EFB=4\times$ area of $\triangle ADE$.

Also, since $\angle DEA=\angle FBE$ and $\angle A$ is a common angle, therefore we have $\triangle ADE \backsim \triangle ACB$.

Similarly, we can obtain

$$\frac{\text{area of }\triangle ADE}{\text{area of }\triangle ACB}=\left(\frac{AE}{AB}\right)^2=\left(\frac{AE}{AE+EB}\right)^2=\left(\frac{1}{3}\right)^2=\frac{1}{9}$$

and therefore area of $\triangle ACB=9\times$ Area of $\triangle ADE$. Thus,

$$\begin{aligned}\text{area of } CDEF &= \text{area of }\triangle ACB-\text{area of }\triangle ADE-\text{area of }\triangle EFB\\ &=(9-1-4)\times\text{area of }\triangle ADE=4\times\text{area of }\triangle ADE\end{aligned}$$

Hence,

$$\frac{\text{area of } CDEF}{\text{area of }\triangle EFB}=\frac{4\times\text{area of }\triangle ADE}{4\times\text{area of }\triangle ADE}=1$$

The answer is (B).

解　析

已知 $EF//AC$，所以 $\angle DAE=\angle FEB$。同理，已知 $DE//BC$，所以 $\angle DEA=\angle FBE$。

因此 $\triangle ADE \backsim \triangle EFB$，则

$$\frac{\triangle ADE\text{ 的面积}}{\triangle EFB\text{ 的面积}}=\left(\frac{AE}{EB}\right)^2=\left(\frac{1}{2}\right)^2=\left(\frac{1}{4}\right)^2$$

因此 $\triangle EFB$ 的面积 $=4\times\triangle ADE$ 的面积。

由于 $\angle DEA=\angle FBE$ 且 $\angle A$ 是公共角，因此 $\triangle ADE \backsim \triangle ACB$。

同理，

$$\frac{\triangle ADE\text{的面积}}{\triangle ACB\text{的面积}}=\left(\frac{AE}{AB}\right)^2=\left(\frac{AE}{AE+EB}\right)^2=\left(\frac{1}{3}\right)^2=\frac{1}{9}$$

因此 $\triangle ACB$ 的面积 $=9\times\triangle ADE$ 的面积。

由此可得

四边形 $CDEF$ 的面积 $=\triangle ACB$ 的面积 $-\triangle ADE$ 的面积 $-\triangle EFB$ 的面积

$=(9-1-4)\times\triangle ADE$ 的面积 $=4\times\triangle ADE$ 的面积

因此

$$\frac{CDEF\text{ 的面积}}{\triangle EFB\text{ 的面积}}=\frac{4\times\triangle ADE\text{ 的面积}}{4\times\triangle ADE\text{ 的面积}}=1$$

故答案是 (B)。

PRACTICE　练习

Problem 1

The diagram shows a seven-sided polygon, *PQRSTUV*. It is formed by two equilateral triangles *PQW* and *STU* of side-length 5cm and 8cm respectively. The two triangles overlap in an equilateral triangle of side-length 2cm. What is the ratio of the area of *PVRQ* to the area of *UVWRST*?

问题 1

图中显示了一个七边形 *PQRSTUV*。它由两个边长分别为 5 厘米和 8 厘米的等边 $\triangle PQW$ 和 $\triangle STU$ 组成。这两个三角形在边长为 2 厘米的等边三角形中重叠。问 *PVRQ* 和 *UVWRST* 的面积之比是多少？

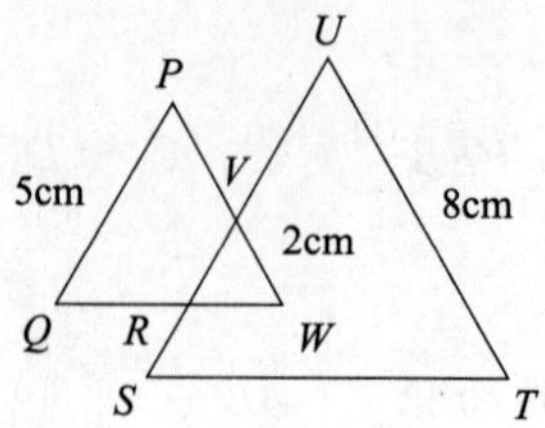

(A) 7 : 20　　(B) 21 : 64　　(C) 25 : 64　　(D) 21 : 60　　(E) 23 : 62

Problem 2

The area of triangle XYZ is 16 square inches. Points A and B are midpoints of congruent segments XY and XZ. Altitude XC bisects YZ. What is the area (in square inches) of the shaded region ?

问题 2

$\triangle XYZ$ 的面积为 16 平方英寸。点 A 和 B 是相等线段 XY 和 XZ 的中点。垂线 XC 平分 YZ。问阴影区域的面积（平方英寸）是多少？

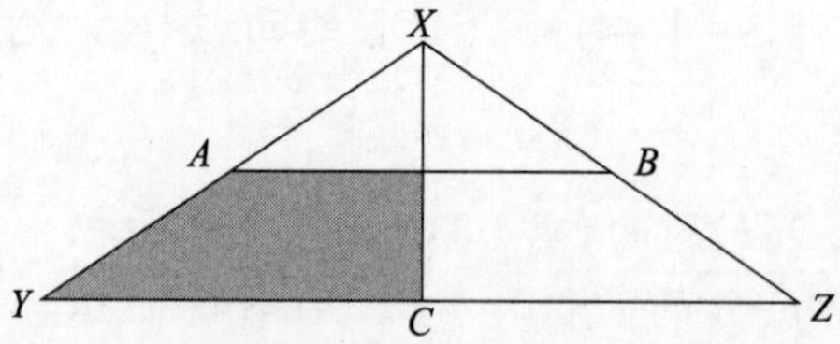

(A) 4　　(B) 5　　(C) 5.5　　(D) 6　　(E) 8

Problem 3

In a right triangle, a perpendicular is dropped from the right angle to the hypotenuse and the segments of the hypotenuse have lengths of x inches and $3x$ inches. If the altitude is 3 inches in length, then what is the length of x in inches?

问题 3

在一个直角三角形中，从直角向斜边做一条垂线并且斜边被分为长度为 x 英寸和 $3x$ 英寸的两条线段。如果高为 3 英寸，那么 x 的长度是多少英寸？

(A) $\frac{1}{3}$　　(B) 1　　(C) $\sqrt{3}$

(D) 3　　(E) cannot be determined 不能确定

Problem 4

What is the length of AC in the right triangle $\triangle ABC$ if $AD \perp BC$?

问题 4

在直角 $\triangle ABC$ 中，如果 $AD \perp BC$，AC 的长度是多少？

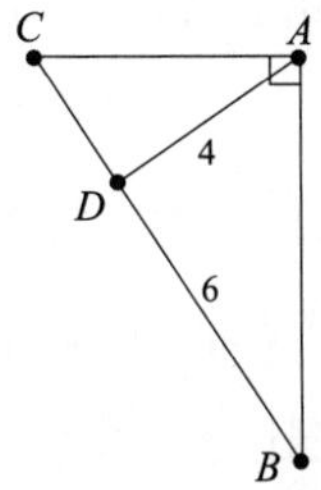

(A) $\frac{26}{3}$ (B) $2\sqrt{13}$ (C) $\frac{10\sqrt{13}}{3}$ (D) $\frac{4\sqrt{13}}{3}$ (E) $\frac{8}{3}$

Problem 5

What is the measure of the acute angle formed by the hands of the clock at 4:40 PM?

问题 5

在下午 4:40，时针与分针形成的锐角是多少度？

(A) 80° (B) 90° (C) 100° (D) 120° (E) 140°

Problem 6

Given the areas of the three squares in the figure, what is the total area of the figure below?

问题 6

给定下图中三个正方形的面积，问整个图形的面积是多少？

(A) 13 (B) 30 (C) 60 (D) 300 (E) 368

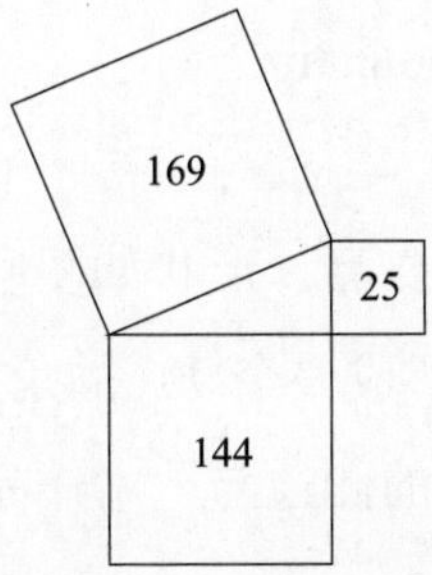

Problem 7

Isosceles triangle ABC has equal side lengths AB and BC. In the figure below, segments are drawn parallel to AC so that the shaded portions of triangle ABC have the same area. The heights of the two unshaded portions are 11 and 5 units, respectively. What is the ratio of the height of the unshaded triangle on the left to h of triangle ABC?

问题 7

等腰△ABC的边长AB和BC相等。在下图中，分别画出两条与AC平行的线段使两图中△ABC的阴影部分面积相等。两个未着色部分的高分别为 11 和 5。请问左图未着色部分的高和△ABC的高 h 之比是多少？

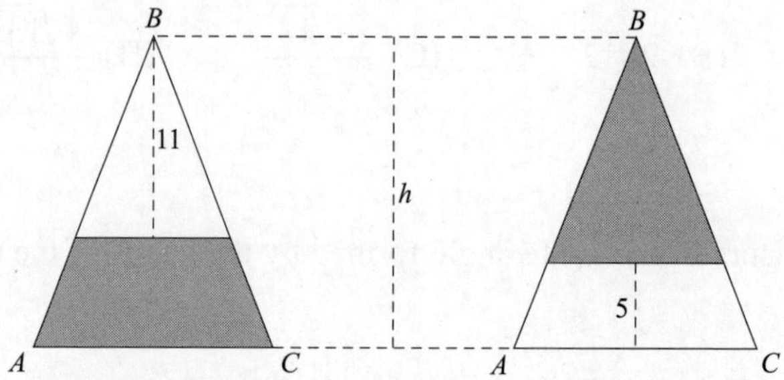

（注：图形未按比例绘制）

(A) 55 : 74　(B) 55 : 73　(C) 11 : 15　(D) 55 : 76　(E) 5 : 7

Problem 8

Jean has made a piece of stained-glass art in the shape of two mountains, as shown in the figure below. One mountain peak is 8 feet high while the other peak is 12 feet high. Each peak forms a 90° angle, and the straight sides form a 45° angle with the ground. The artwork has an area of 183 square feet. The sides of the mountain meet at an intersection point near the center of the artwork, h feet above the ground. What is the the ratio of the value of h to the value of the lower height of the mountain?

问题 8

吉恩制作了一件彩色玻璃艺术作品，形状如图所示的两座山。一个山峰高 8 英尺，另一个山峰高 12 英尺。每个山峰形成一个 90° 的角，直边与地面形成 45° 的角。该艺术作品的面积为 183 平方英尺。山的侧边与艺术作品中心附近的交点高于地面 h 英尺。问 h 和较矮的山峰的高之比为多少？

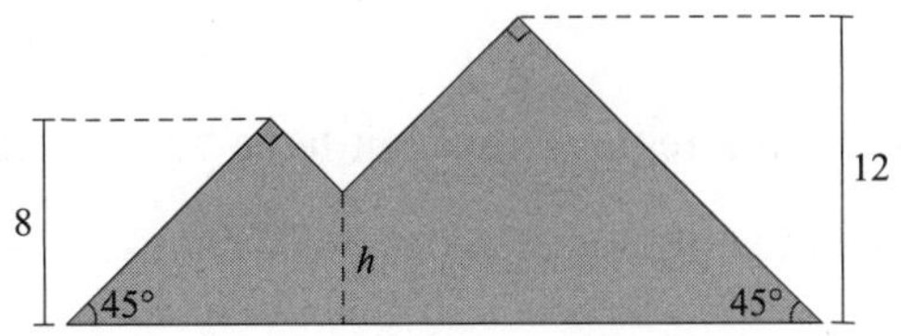

(A) 1 : 2 (B) 5 : 8 (C) $5\sqrt{2}:12$ (D) $5\sqrt{2}:8$ (E) 3 : 4

Problem 9

The lengths of the sides of a triangle in inches are three consecutive integers. The length of the shortest side is 25% of the perimeter. What is the length of the longest side?

问题 9

以英寸为单位的一个三角形的边长是 3 个连续的整数。最短边的长度是周长的 25%。最长边的长度是多少？

(A) 3 (B) 4 (C) 5 (D) 6 (E) 7

Problem 10

In $\triangle ABC$, AB=BC=26, and AC=48. What is the area of $\triangle ABC$?

问题 10

在 $\triangle ABC$ 中，$AB=BC=26$，$AC=48$。求三角形 ABC 的面积。

(A) 168 (B) 192 (C) 240 (D) 264 (E) 288

Problem 11

In the figure below, choose point D on BC so that $\triangle ACD$ and $\triangle ABD$ have equal perimeters. What is the area of $\triangle ADC$?

问题 11

在下图中，在 BC 上选一点 D，使得 $\triangle ACD$ 和 $\triangle ABD$ 的周长相等。问 $\triangle ADC$ 的面积是多少？

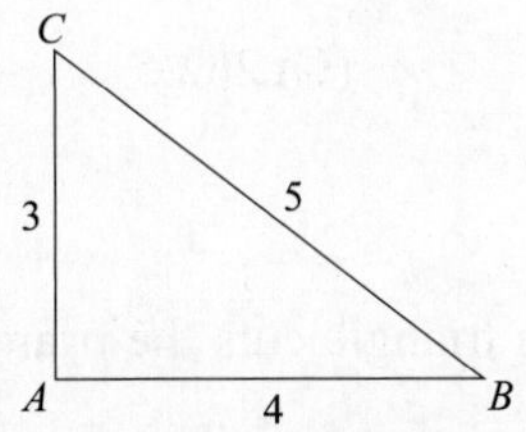

(A) $\dfrac{3}{2}$ (B) 2 (C) $\dfrac{12}{5}$ (D) $\dfrac{5}{2}$ (E) $\dfrac{18}{5}$

Problem 12

An equilateral triangle and a regular hexagon have equal perimeters. If the area of the triangle is 8, then what is the area of the hexagon?

问题 12

一个等边三角形和一个正六边形的周长相等。如果三角形的面积是 8，那么六边形的面积是多少?

(A) 8 (B) 10 (C) 12 (D) $8\sqrt{3}$ (E) $12\sqrt{3}$

Problem 13

If $AB//DE$, $AB = 5$, $CE = 8$, and $DE = 12$, find the measure of BC.

问题 13

如果 $AB//DE$，$AB = 5$，$CE = 8$，$DE = 12$，求 BC 的长度。

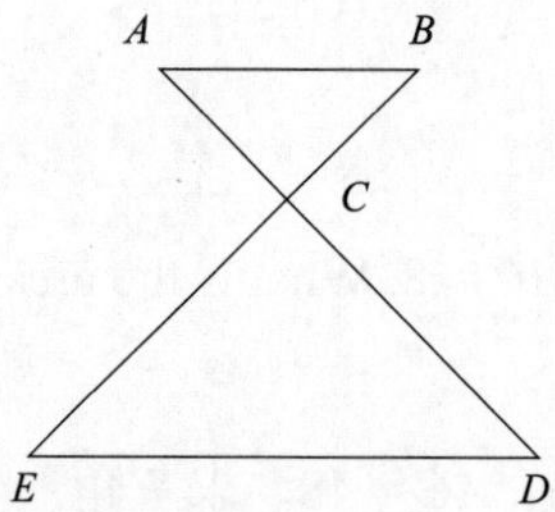

(A) 3 (B) 7.5 (C) 15 (D) $\dfrac{10}{3}$ (E) 4

Problem 14

A triangle with sides 9, 12 and 15 is similar to another triangle which has longest side 35. The area of the larger triangle is:

问题 14

一个边长为 9，12 和 15 的三角形与另一个最长边为 35 的三角形相似。问较大三角形的面积是多少?

(A) 54 (B) 216 (C) 262.5 (D) 294 (E) 588

Problem 15

A line parallel to the base of a triangle cuts the triangle into two regions, with the ratio of the smaller area to the bigger area being 1 : 2. This line also cuts the altitude into two parts. Find the ratio of the two parts of the altitude.

问题 15

一条平行于三角形底边的线将三角形分成两个区域，小三角形和大三角形的面积之比为 1 : 2。该线也将高分成两部分。求这两部分线段的长度比。

(A) 1 : 1 (B) 1 : 2 (C) $1 : \sqrt{2}$

(D) $1 : \sqrt{2} - 1$ (E) none of these 以上都不是

Chapter 9 Polygons

第 9 章 多边形

9.1 Basic Knowledge 基础内容

9.1.1 Squares 正方形

Area of a square: $A = a^2$, where a is the side length.

Length of diagonal: $d = \sqrt{2}a$, where a is the side length.

Perimeter of a square: $P = 4a$, where a is the side length.

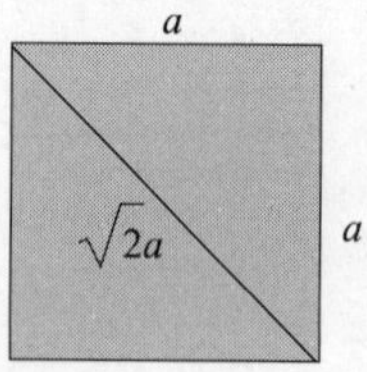

正方形面积：$A = a^2$，其中 a 是边长。

对角线长度：$d = \sqrt{2}a$，其中 a 是边长。

正方形周长：$P = 4a$，其中 a 是边长。

Example 1

Four squares of side length 4, 7, 9 and 10 are arranged in increasing size order so that their left edges and bottom edges align. The squares alternate in color white-gray-white-gray, respectively, as shown in the figure. What is the area of the visible gray region in square units?

例题 1

边长为 4、7、9 和 10 的 4 个正方形按大小递增的顺序排列，它们的左边缘和底边缘对齐。如图所示，正方形的颜色分别为白、灰、白、灰。问可见灰色区域的面积是多少？

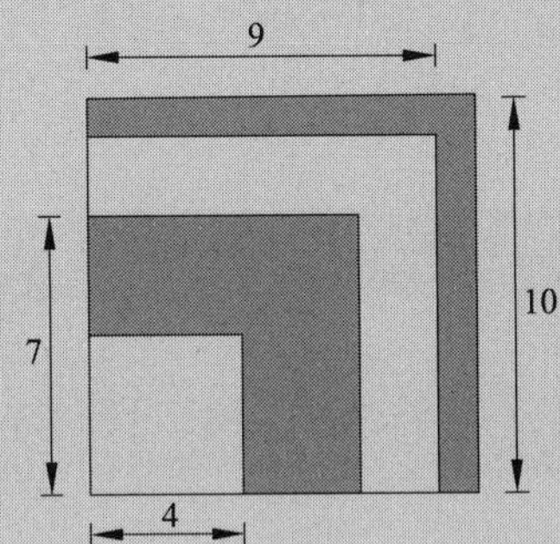

(A) 42 (B) 45 (C) 49 (D) 50 (E) 52

Solution

We work inwards. The area of the outer shaded square is difference of the area of the whole square and the area of the second largest square. The area of the inner shaded region is the difference of area of the third largest square and the area of the smallest square. The area we want is:

$$10^2 - 9^2 + 7^2 - 4^2 = 19 + 33 = \text{(E) } 52$$

解 析

由图可得，外阴影部分的面积等于整个正方形的面积和第二大正方形的面积的差。内阴影区域的面积是第三大正方形的面积和最小正方形的面积差。我们所求的面积之和就是 $10^2 - 9^2 + 7^2 - 4^2 = 19 + 33 = 52$，故选 (E)。

9.1.2 Rectangle 长方形（矩形）

Area of a rectangle: $A = b \cdot h$.

Length of diagonal: $d = \sqrt{b^2 + h^2}$.

Perimeter of a rectangle: $P = 2(b + h)$.

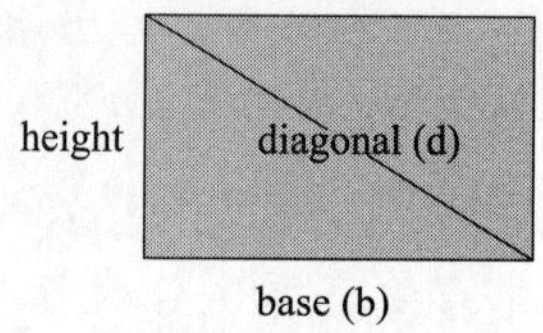

矩形面积：$A = b \cdot h$。

对角线长度：$d=\sqrt{b^2+h^2}$。

矩形周长：$P=2(b+h)$。

Example 2

The area of polygon $ABCDEF$ is 56, with $AB=8$, $BC=9$ and $FA=5$. What is $DE+EF$?

例题 2

多边形 $ABCDEF$ 的面积是 56，其中 $AB=8$，$BC=9$，$FA=5$。求 $DE+EF$ 是多少？

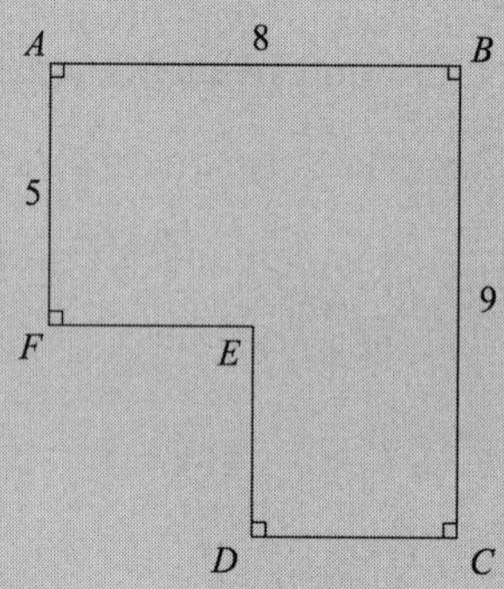

(A) 7 (B) 8 (C) 9 (D) 10 (E) 11

Solution

Notice that $AF+DE=BC$, so $DE=4$. Let O be the intersection of the extensions of AF and DC, which makes rectangle $ABCO$. The area of the polygon equals the difference between the area of $FEDO$ and the area of $ABCO$. Thus we have $56=8\times 9-EF\times 4$. Solving for the unknown, $EF=4$, therefore $DE + EF=4+4=$ (B)8.

解 析

因为 $AF+DE=BC$，所以 $DE=4$。设 O 是 AF 和 DC 的延长线的交点，得到矩形 $ABCO$。多边形的面积是 $FEDO$ 的面积和 $ABCO$ 的面积之差，所以 $56=8\times 9-EF\times 4$。求出未知数 $EF=4$，因此 $DE+EF=4+4=8$，故选 (B)。

Example 3

The length of a rectangle is increased by 20% and the width is decreased by 20%. What percent of the old area is the new area?

例题 3

矩形的长增加 20%，宽减少 20%。问新面积是旧面积的百分之多少？

(A) 90%　　(B) 96%　　(C) 100%　　(D) 104%　　(E) 110%

Solution

In a rectangle with dimensions 10×10, the new rectangle would have dimensions 12 and 8. So, the ratio of the new area to the old area is 96/100 = 96%. The answer is (B).

More generally, let a and b be the length and width of the rectangle respectively. Then the original area is ab and the new area is $1.2a\times 0.8b=0.96ab$. Therefore, the ratio is $\frac{0.96ab}{ab}\times 100\%=96\%$.

解 析

在一个尺寸为 10×10 的矩形中，新矩形的尺寸为 12×8。新面积与旧面积的比值是 96/100 = 96%，故选 (B)。

更一般地，可设长为 a，宽为 b，则原矩形面积为 ab，新矩形面积为 $1.2a\times 0.8b=0.96ab$，新面积与旧面积的比值为 $\frac{0.96ab}{ab}=96\%$。

Example 4

Rectangles R_1 and R_2, and squares S_1, S_2 and S_3, shown below, combine to form a rectangle that is 920 units wide and 680 units high. What is the side length of S_2 in units?

例题 4

如下图所示，矩形 R_1 和 R_2 以及正方形 S_1，S_2 和 S_3 组合在一起形成一个长 920，宽 680 的矩形。问正方形 S_2 的边长是多少？

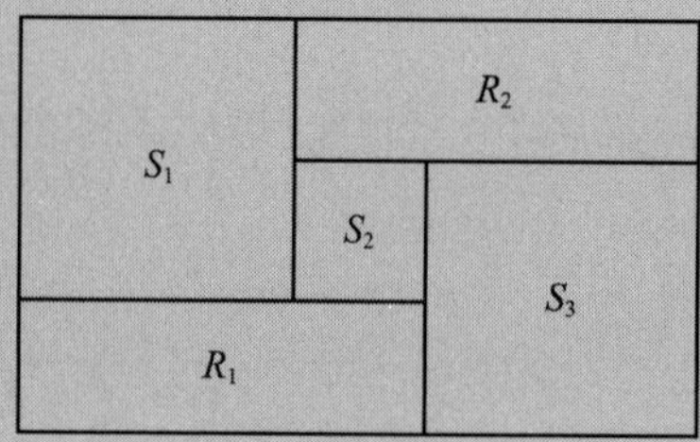

(A) 120 (B) 124 (C) 126 (D) 130 (E) 140

Solution

Let the side length of each square S_k be s_k. Then, from the diagram, we have $s_1+s_2+s_3=920$, and $s_1-s_2+s_3=680$. We subtract the second equation from the first to obtain $2s_2 = 240$, and thus $s_2 =$ (A) 120.

解 析

设每个正方形 S_k 的边长为 s_k。那么，从图中可以得出 $s_1+s_2+s_3=920$，$s_1-s_2+s_3=680$。将第二个方程减去第一个方程，得到 $2s_2=240$，因此 $s_2=120$，故选 (A)。

9.1.3 Parallelogram 平行四边形

Area of parallelogram: $A = b \cdot h$, where b is the base and h is the height.

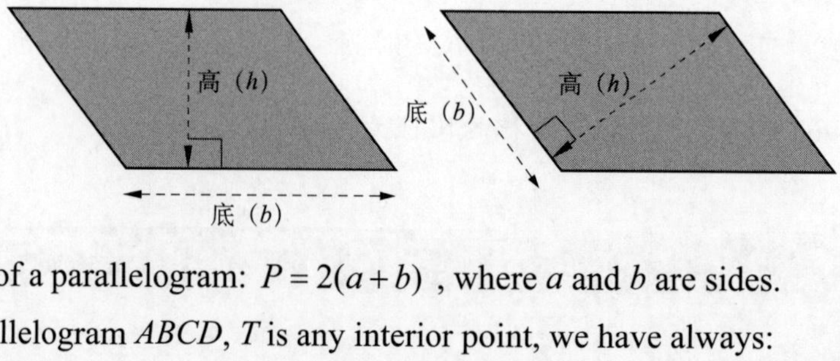

perimeter of a parallelogram: $P = 2(a+b)$, where a and b are sides.

In the parallelogram $ABCD$, T is any interior point, we have always:

$$S_{\triangle TAB} + S_{\triangle TCD} = \frac{1}{2} \times S_{ABCD}$$

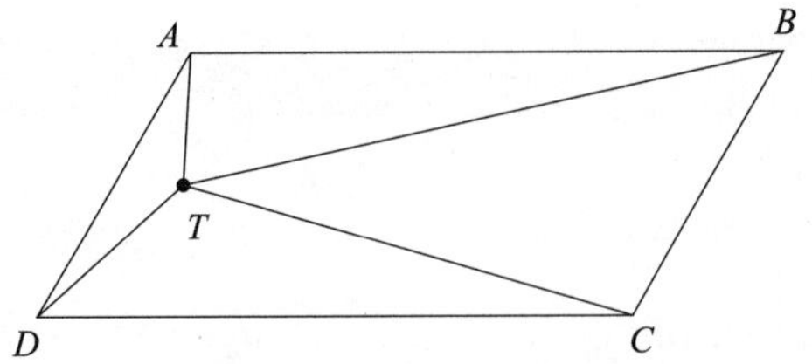

平行四边形面积：$A = b \cdot h$，其中 b 是底，h 是高。

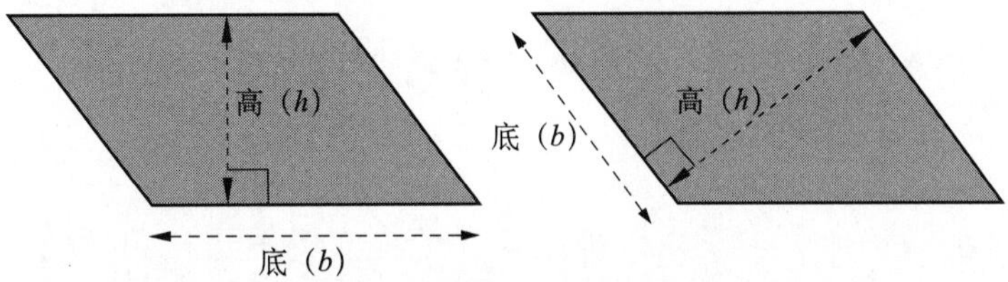

平行四边形周长：$P = 2(a+b)$，a 是另一条边的长度。

在平行四边形 $ABCD$ 中，T 是任意一个内点，总有

$$S_{\triangle TAB} + S_{\triangle TCD} = \frac{1}{2} \times S_{ABCD}$$

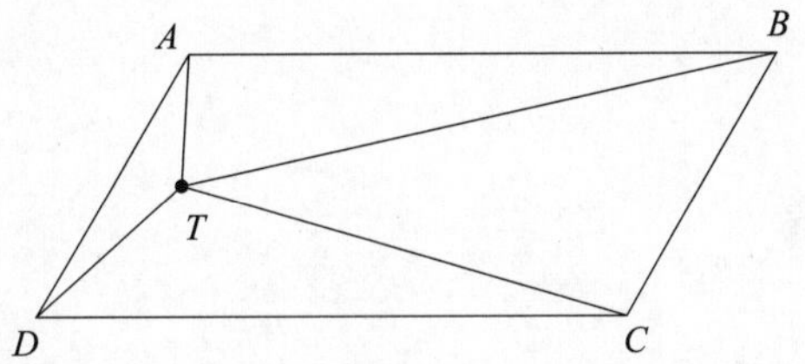

9.1.4 Trapezoid 梯形

Area of a trapezoid: $A = \frac{(b_1 + b_2)}{2}h$.

This can be thought as the area of two triangles added together.

梯形面积：$A = \frac{(b_1 + b_2)}{2}h$。

这相当于两个三角形的面积之和。

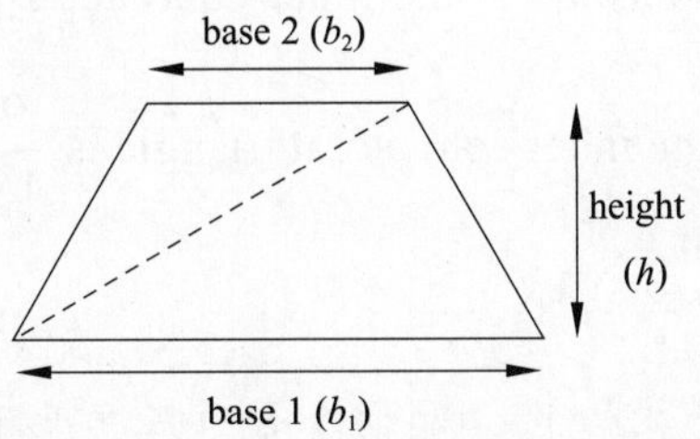

Example 5

An equilateral triangle is placed inside a larger equilateral triangle so that the region between them can be divided into three congruent trapezoids, as shown below. The side length of the inner triangle is $\frac{4}{5}$ the side length of the larger triangle. What is the ratio of the area of one trapezoid to the area of the inner triangle?

例题 5

如下图所示，一个等边三角形被放在一个更大的等边三角形内，这样它们之间的区域就被分成 3 个全等的梯形。内三角形的边长是大三角形边长的 $\frac{4}{5}$。问一个梯形的面积与内部三角形的面积之比是多少？

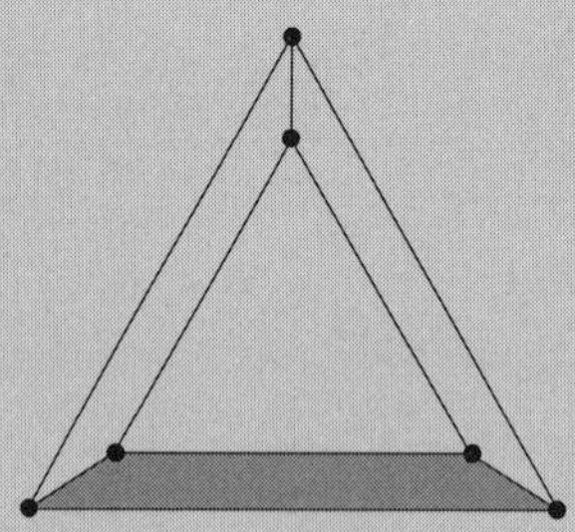

(A) 1 : 3　　(B) 3 : 8　　(C) 3 : 16　　(D) 7 : 16　　(E) 4 : 9

Solution

All equilateral triangles are similar. Since ratio of the side lengths of the outer equilateral triangle to the inner equilateral triangle is $\frac{5}{4}$, the ratio of their areas is $\left(\frac{5}{4}\right)^2=\frac{25}{16}$. It follows that the area ratio of three trapezoids to the inner equilateral triangle is $\frac{25}{16}-1=\frac{9}{16}$, so the area ratio of one trapezoid to the inner equilateral triangle is $\frac{9}{16}\times\frac{1}{3}=$ (C) $\frac{3}{16}$.

解　析

所有的等边三角形都是相似的。对于外等边三角形和内等边三角形来说，因为它们的边长比是 $\frac{5}{4}$，所以面积比是 $\left(\frac{5}{4}\right)^2=\frac{25}{16}$。所以三个梯形的面积与内等边三角

形的面积比为 $\frac{25}{16}-1=\frac{9}{16}$，所以一个梯形与内等边三角形的面积比为 $\frac{9}{16}\times\frac{1}{3}=\frac{3}{16}$，故选 (C)。

9.1.5 Rhombus 菱形

Perimeter of a rhombus: $P=4a$, where a is the side length.

Area of a rhombus: $A=\frac{1}{2}d_1\times d_2$, where d_1 and d_2 are diagonals.

菱形周长：$P=4a$, 其中 a 是边长。

菱形面积：$A=\frac{1}{2}d_1\times d_2$, 其中 d_1 和 d_2 是对角线。

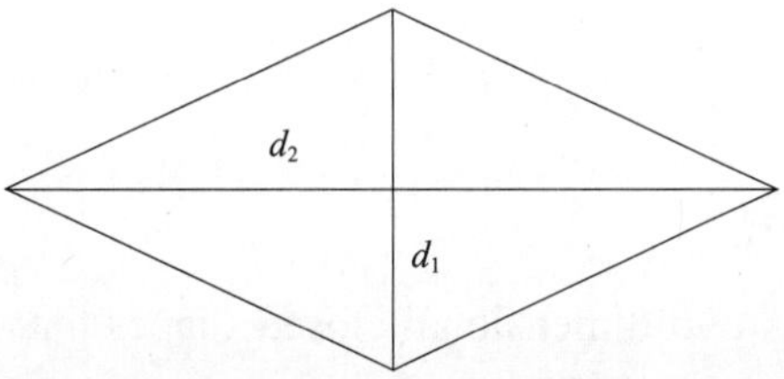

Example 6

Points A, B, C and D are midpoints of the sides of the larger square. If the larger square has area 40, then what is the area of the smaller square?

例题 6

点 A、B、C 和 D 是大正方形边的中点。如果大正方形的面积是 40，那么小正方形的面积是多少？

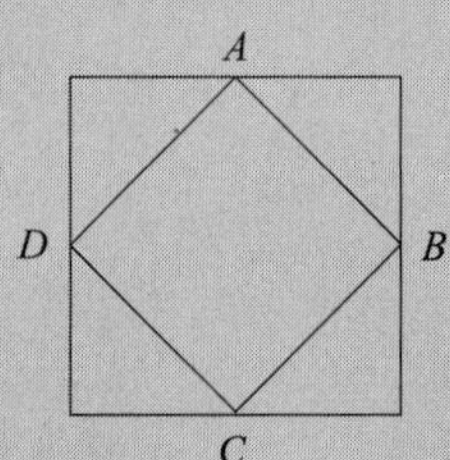

(A) 15　　(B) 20　　(C) 24　　(D) 30　　(E) 40

Solution

Drawing segments AC and BD, we can divide the larger square into 8 congruent right isosceles triangles. Since the number of triangles outside square $ABCD$ is the same as the number of triangles inside the square, the areas inside and outside $ABCD$ must be equal. So the area of $ABCD$ is half of the area of the larger square, and equals to $\frac{40}{2}=$ (B) 20.

解　析

画出线段 AC 和 BD，则大正方形被分割为 8 个全等的等腰直角三角形。由于正方形 $ABCD$ 外的三角形个数与正方形内的三角形数相同，$ABCD$ 内外的面积必须相等。所以 $ABCD$ 的面积是较大正方形面积的一半，即 $\frac{40}{2}=20$，故选（B）。

9.1.6　Polygons 多边形

Polygons are defined as two-dimensional closed shapes that are formed by joining three or more-line segments with each other.

多边形是由三条或三条以上线段相互连接而成的二维封闭图形。

Names of Polygons:

认识多边形：

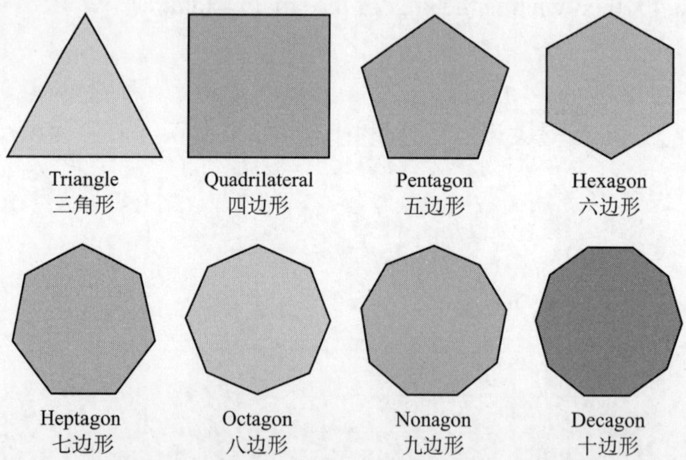

Example 7

In the given figure, hexagon $ABCDEF$ is equiangular, $ABJI$ and $FEHG$ are squares with areas 18 and 32 respectively, $\triangle JBK$ is equilateral and $FE=BC$. What is the area of $\triangle KBC$?

例题 7

在给定的图形中，六边形 $ABCDEF$ 中六个内角相等，$ABJI$ 和 $FEHG$ 分别是面积为 18 和 32 的正方形，$\triangle JBK$ 是等边三角形，$FE=BC$。$\triangle KBC$ 的面积是多少？

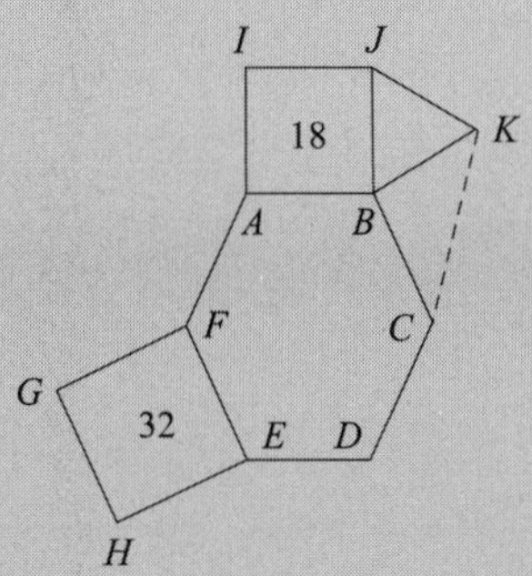

(A) $6\sqrt{2}$　　(B) 9　　(C) 12　　(D) $9\sqrt{2}$　　(E) 32

Solution

Since FE is a side of a square with area 32, $FE=\sqrt{32}=4\sqrt{2}$. Now, since $FE=BC$, we have $BC=4\sqrt{2}$. Now, JB is a side of a square with area 18, so $JB=\sqrt{18}=3\sqrt{2}$. Since $\triangle JBK$ is equilateral, $BK=3\sqrt{2}$. Since $ABCDEF$ is equiangular, we know $\angle ABC=120°$. Finally, we see that $\angle CBK=360°-(\angle JBA+\angle ABC+\angle KBJ)=360°-(90°+120°+60°)=90°$. So $\triangle KBC$ is a right triangle with legs $3\sqrt{2}$ and $4\sqrt{2}$. Thus, its area is $\frac{3\sqrt{2}\times 4\sqrt{2}}{2}=\frac{24}{2}=\text{(C)}12$.

解　析

因为 FE 是面积为 32 的正方形的边，所以 $FE=\sqrt{32}=4\sqrt{2}$。因为 $FE=BC$，所以 $BC=4\sqrt{2}$。由于 JB 是面积为 18 的正方形的边，所以 $JB=\sqrt{18}=3\sqrt{2}$。因为 $\triangle JBK$ 是等边三角形，所以 $BK=3\sqrt{2}$。又因为 $ABCDEF$ 中 6 个内角相等，我们可知 $\angle ABC=120°$。因此 $\angle CBK=360°-(\angle JBA+\angle ABC+\angle KBJ)=360°-(90°+120°+60°)=90°$。所以 $\triangle KBC$ 是一个直角三角形，它的两条直角边分别是 $3\sqrt{2}$ 和 $4\sqrt{2}$。则它的面积为 $\frac{3\sqrt{2}\times 4\sqrt{2}}{2}=\frac{24}{2}=12$，故选 (C)。

Example 8

The diagram shows an octagon consisting of 10-unit squares. The portion below PQ is a unit square and a triangle with base 5. If PQ bisects the area of the octagon, what is the ratio $\dfrac{YQ}{QX}$?

例题 8

图中显示的是一个由 10 个单位正方形组成的八边形。PQ 下面的部分是一个单位正方形和一个底为 5 的三角形。如果 PQ 平分八边形的面积，那么比值 $\dfrac{YQ}{QX}$ 是多少?

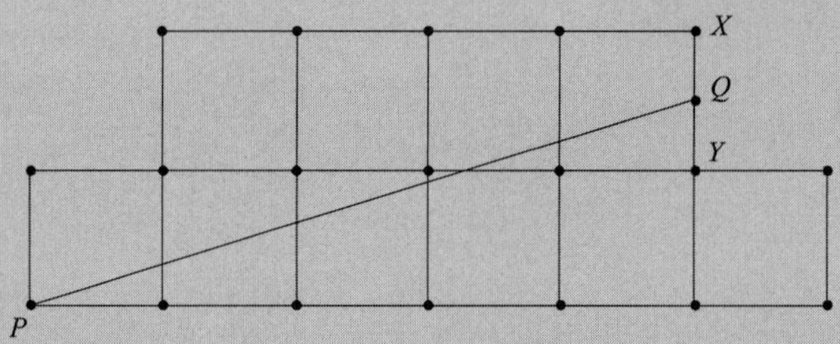

(A) $\dfrac{2}{3}$　(B) $\dfrac{5}{4}$　(C) 1　(D) $\dfrac{3}{2}$　(E) $\dfrac{3}{4}$

Solution

Since the area of the octagon is 10, we know that half of the area of the octagon is 5. Thus the area of the right triangle is $5-1=4$. Let the verticle height of the triangle be h. Then we have $\dfrac{5h}{2}=4$ and $h=\dfrac{8}{5}$, so $QY=\dfrac{8}{5}-1=\dfrac{3}{5}$, and $XQ=1-\dfrac{3}{5}=\dfrac{2}{5}$. Therefore, the ratio is $\dfrac{\frac{3}{5}}{\frac{2}{5}}=$ (D) $\dfrac{3}{2}$.

解　析

由于八边形的面积是 10，则面积的一半是 5。所以三角形的面积是 $5-1=4$。设三角形的高为 h，则有 $\dfrac{5h}{2}=4$，解得 $h=\dfrac{8}{5}$，因此 $QY=\dfrac{8}{5}-1=\dfrac{3}{5}$，且 $XQ=1-\dfrac{3}{5}=\dfrac{2}{5}$。故

比值为 $\dfrac{\frac{3}{5}}{\frac{2}{5}}=\dfrac{3}{2}$，故选 (D)。

Example 9

Point O is the center of the regular octagon $ABCDEFGH$, and X is the midpoint of the side AB. What fraction of the area of the octagon is shaded?

例题 9

点 O 是正八边形 $ABCDEFGH$ 的中心，X 是边 AB 的中点。问阴影部分占八边形面积的几分之几？

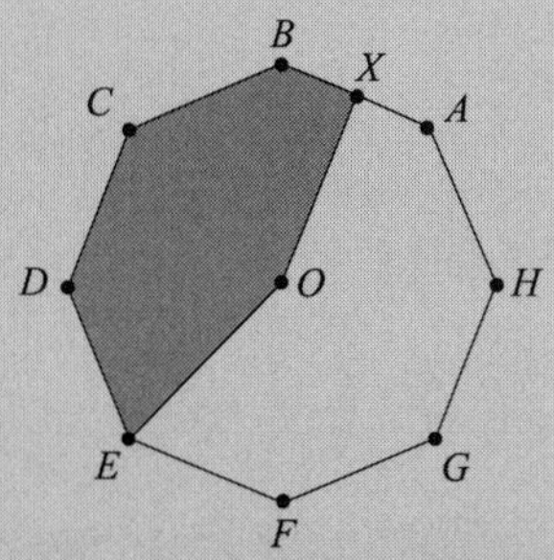

(A) $\dfrac{11}{32}$ (B) $\dfrac{3}{8}$ (C) $\dfrac{13}{32}$ (D) $\dfrac{7}{16}$ (E) $\dfrac{15}{32}$

Solution

As shown in the diagram, the octagon has been divided up into 16 identical triangles (and thus they each have equal area). Since the shaded region occupies 7 out of the 16 total triangles, the answer is (D) $\dfrac{7}{16}$.

解 析

如图所示，正八边形被分割成16个相同的三角形（因此每个三角形的面积相等）。由于阴影部分占了16个三角形中的7个，所以答案是 $\dfrac{7}{16}$，故选 (D)。

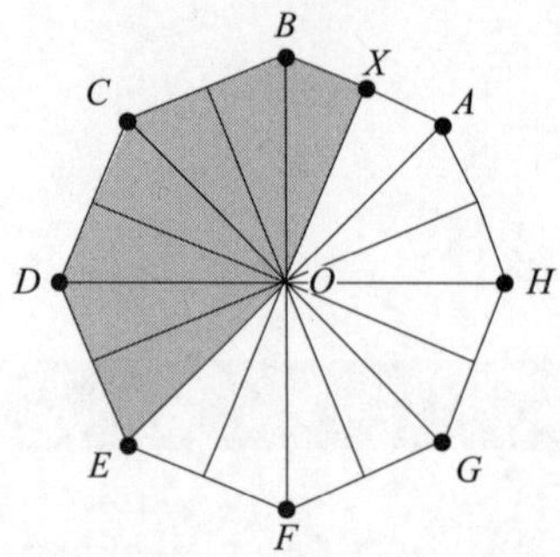

9.1.7 The Pick's Law (Finding the area of the region bounded by grids) 皮克定理（求网格区域的面积）

In a rectangular grid, the area of a simple polygon whose vertices are lattice points is area= $\frac{B}{2}+I-1$.

B: Number of boundary points.

I: Number of interior points.

在一个长方形网格中，一个顶点在格点上的简单多边形满足：面积 = $\frac{B}{2}+I-1$，其中 B 为边界点数量，I 为内部点数量。

Example 10

Suppose the horizontal and vertical distances between adjacent points equal 1 unit. What is the area of triangle ABC ?

例题 10

假设相邻点之间的水平距离和垂直距离都等于 1 个单位。问 $\triangle ABC$ 的面积是多少?

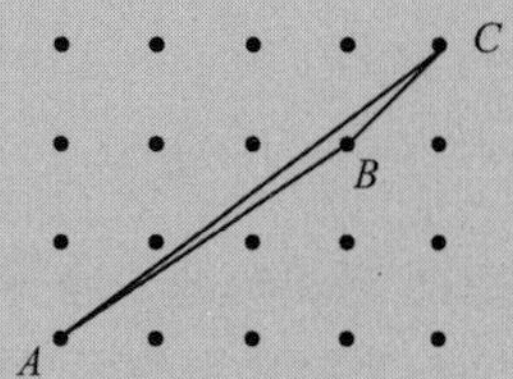

(A) $\frac{1}{4}$ (B) $\frac{1}{2}$ (C) $\frac{3}{4}$ (D) 1 (E) $\frac{5}{4}$

Solution

By Pick's Law, we can simply use the number of coordinate points to get the area. In application, $B = 3, I = 0$, so we get the area $\frac{3}{2} + 0 - 1 = \text{(B)} \frac{1}{2}$.

解 析

根据皮克定律，我们可以简单地使用坐标点的数量来求得面积。由图可知，$B = 3$，$I = 0$，因此我们可以得到面积$\frac{3}{2} + 0 - 1 = \frac{1}{2}$，故选 (B)。

PRACTICE 练习

Problem 1

In the figure, the outer equilateral triangle has area 22, the inner equilateral triangle has area 1, and the three trapezoids are congruent. What is the area of one of the trapezoids?

问题 1

在图中，外等边三角形的面积是 22，内等边三角形的面积是 1，三个梯形全等。问其中一个梯形的面积是多少？

(A) 5　　(B) 6　　(C) 7　　(D) 8　　(E) 9

Problem 2

In the diagram, all angles are right angles and the lengths of the sides are given in centimeters. Note the diagram is not drawn to scale. What is the length in X, in centimeters?

问题 2

在图中，所有的角都是直角，边长的单位是厘米。注意该图没有按比例绘制。问 X 的长度是多少厘米？

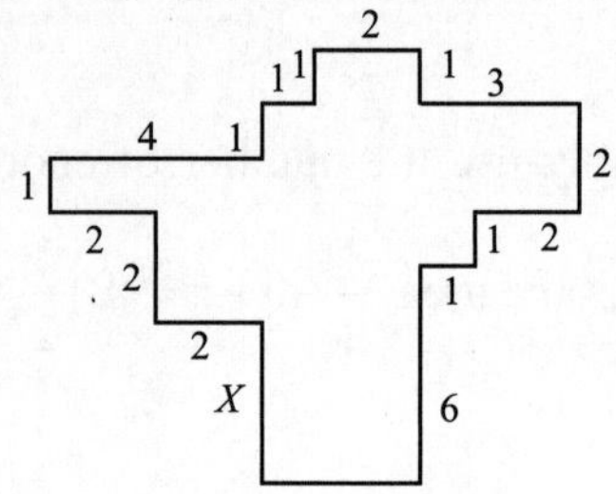

(A) 1 (B) 2 (C) 3 (D) 4 (E) 5

Problem 3

Which of the following polygons has the largest area?

问题 3

下面哪个多边形的面积最大?

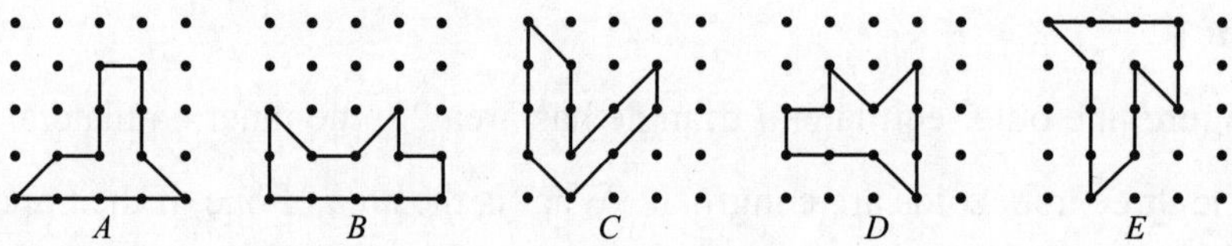

(A) A (B) B (C) C (D) D (E) E

Problem 4

The midpoints of the four sides of a rectangle are $(-2, -1)$, $(3, -1)$, $(6, 3)$and $(1, 3)$. What is the area of the rectangle?

问题 4

矩形 4 条边的中点分别是 $(-2, -1)$, $(3, -1)$, $(6, 3)$ 和 $(1, 3)$。问矩形的面积是多少?

(A) 20 (B) 25 (C) 40 (D) 50 (E) 80

Problem 5

In the figure, the area of square $WXYZ$ is 36cm^2. The four smaller squares have sides 1cm long, either parallel to or coinciding with the sides of the large square. In $\triangle ABC$, $AB=AC$, and when $\triangle ABC$ is folded over side BC, point A coincides with O, the center of square $WXYZ$. What is the area of $\triangle ABC$, in square centimeters?

问题 5

在图中,正方形 $WXYZ$ 的面积是 36 平方厘米。4 个小正方形的边长都是 1 厘米,与大正方形的边平行或重合。在 $\triangle ABC$ 中，$AB=AC$，当把 $\triangle ABC$ 折叠到边 BC 上时，点 A 与正方形 $WXYZ$ 的中心 O 重合。问 $\triangle ABC$ 的面积是多少平方厘米？

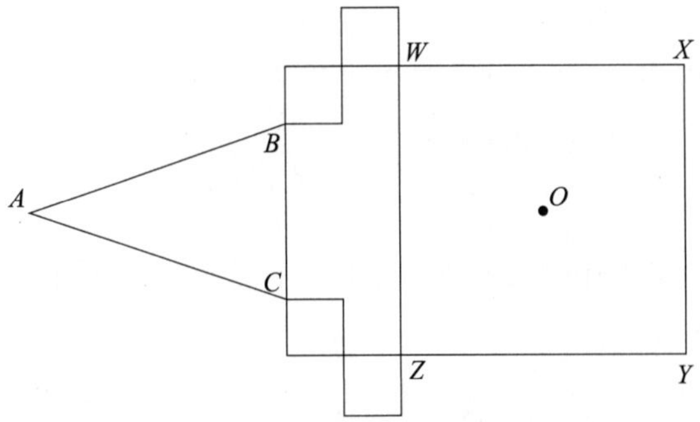

(A)6 (B)8 (C)10 (D)12 (E) 14

Problem 6

In square $ABCE$, $AF : FE = 3 : 2$ and $CD : DE = 3 : 2$. What is the ratio of the area of $\triangle BFD$ to the area of square $ABCE$?

问题 6

在正方形 $ABCE$ 中，$AF : FE = 3 : 2$，$CD : DE = 3 : 2$。问 $\triangle BDF$ 的面积与正方形 $ABCE$ 的面积的比是多少？

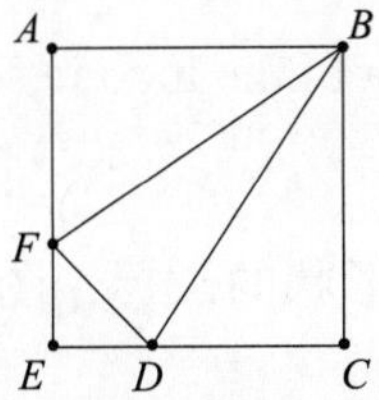

(A) $\frac{9}{25}$ (B) $\frac{6}{25}$ (C) $\frac{8}{25}$ (D) $\frac{2}{5}$ (E) $\frac{3}{5}$

Problem 7

What is the area of the shaded part shown in the 5×5 grid?

问题 7

计算下面 5×5 方格中阴影部分的面积是多少？

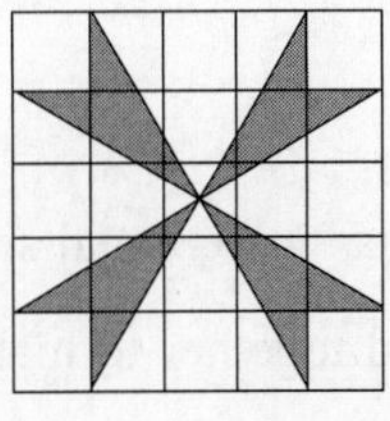

(A) 4 (B) 6 (C) 8 (D) 10 (E) 12

Problem 8

The twelve-sided figure shown has been drawn on 1cm × 1cm graph paper.

What is the area of the figure in cm^2?

问题 8

图中的十二边形是在单位面积为 1 厘米 × 1 厘米的图形纸上画出来的。问这个图形的面积是多少平方厘米?

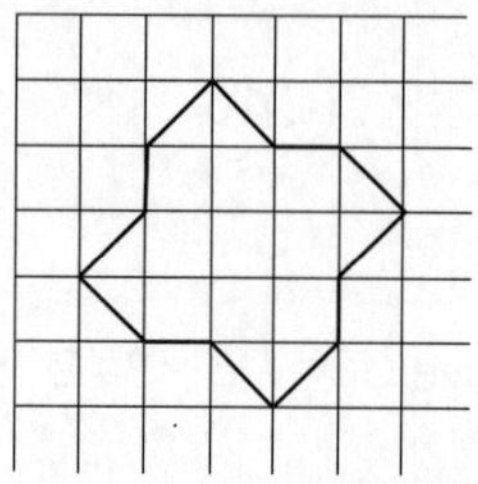

(A) 12 (B) 12.5 (C) 13 (D) 13.5 (E) 14

Problem 9

The Math Team designed a logo shaped like a multiplication symbol, shown below on a grid of 1-inch squares. What is the area of the logo in square inches?

问题 9

数学小组设计了一个乘法符号形状的徽标，图中每个单元格的边长为 1 英寸。求该徽标的面积是多少平方英寸?

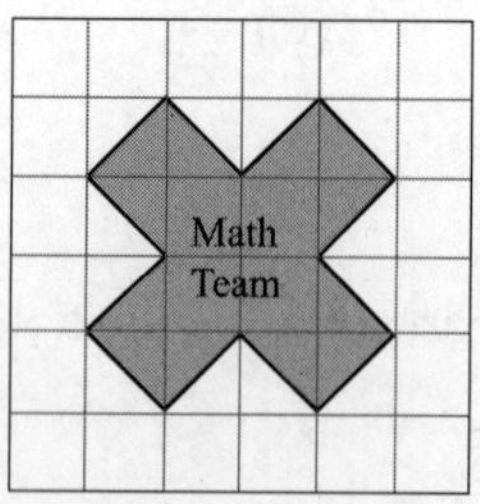

(A) 10 (B) 12 (C) 13 (D) 14 (E) 15

Problem 10

The small trapezium on the left has three equal sides and angles of 60° and 120°. Nine copies of this trapezium can be placed together to make a larger version of it, as shown on the right. The larger trapezium has perimeter 24cm. What is the perimeter of the smaller trapezium?

问题 10

左边的小梯形有 3 条相等的边，其内角分别为 60° 和 120°。9 个小梯形可以拼成一个大梯形。大梯形的周长为 24 厘米。问小梯形的周长是多少？

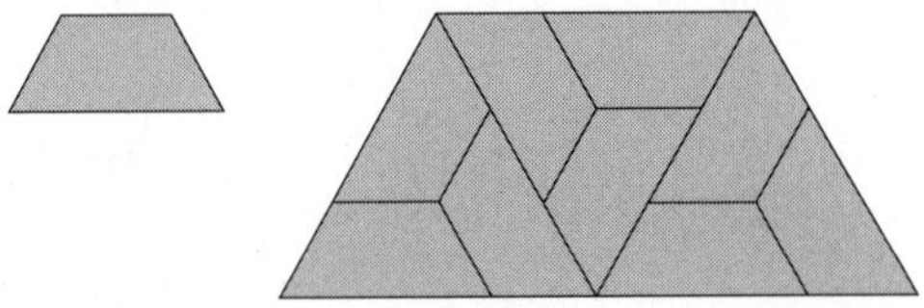

(A) 4cm (B) 6cm (C) 8cm (D) 9cm (E) 10cm

Problem 11

Bill walks $\frac{2}{3}$ mile south, then $\frac{1}{2}$ mile east, and finally $\frac{1}{3}$ mile south. How many miles is he, in a direct line, from his starting point?

问题 11

比尔向南走了 $\frac{2}{3}$ 英里，然后向东走了 $\frac{1}{2}$ 英里，最后向南走了 $\frac{1}{3}$ 英里。从他的出发点算起，问直线距离是多少英里？

(A) $\frac{\sqrt{3}}{2}$ (B) $\frac{\sqrt{5}}{2}$ (C) 1 (D) $\frac{3}{2}$ (E) 2

Problem 12

Construct a square on one side of an equilateral triangle. On one non-adjacent side of the square, construct a regular pentagon, as shown. On a non-adjacent side of the pentagon, construct a hexagon. Continue to construct regular polygons in the same way, until you construct a decagon. How many sides does the resulting polygon have?

问题 12

在等边三角形的一边上拼出一个正方形。在正方形的一条不与之前边相邻的边上，拼出一个正五边形，如图所示。在五边形不与之前边相邻的一边上，拼出一个正六边形。继续用同样的方法拼出正多边形，直到拼出一个正十边形。问这样得到的多边形有几条边？

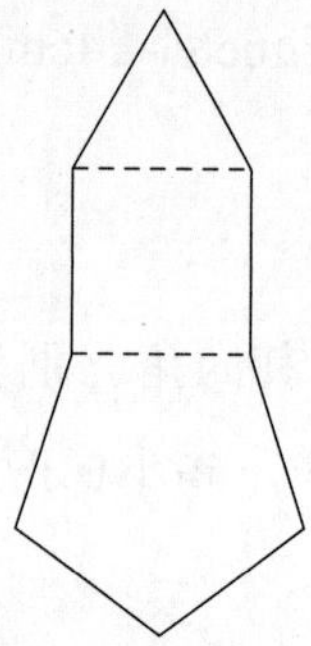

(A) 37 (B) 38 (C) 39 (D) 40 (E) 41

Problem 13

The diagram shows four congruent rectangles, each of perimeter 32cm, surrounding a square of area 44cm^2. What is the area of each rectangle?

问题 13

图中有 4 个全等的矩形，每个矩形的周长都是 32 厘米，围绕着一个面积为 44 平方厘米的正方形。问每个矩形的面积是多少？

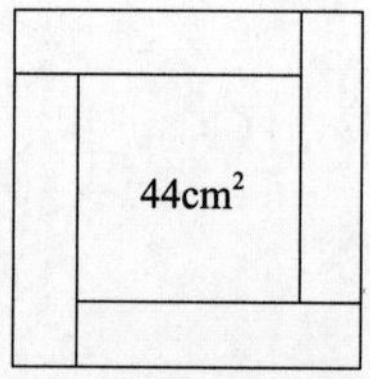

(A) 53cm^2 (B) 55cm^2 (C) 57cm^2 (D) 59cm^2 (E) 61cm^2

Problem 14

Harriet has a square piece of paper. She folds it in half to form a rectangle and then in half again to form a second rectangle (which is not a square). The perimeter of the second rectangle is 40 cm. What is the area of the original square?

问题 14

哈丽特有一张正方形纸。她把它对折成一个长方形，然后再对折成第二个长方形（不是正方形）。第二个长方形的周长是 40 厘米。问原正方形的面积是多少？

(A) 81cm^2 (B) 144cm^2 (C) 196cm^2 (D) 225cm^2 (E) 256cm^2

Problem 15

The diagram shows two squares *JKLM* and *PQRS*. The length of *JK* is 8 cm and that of PQ is 6 cm. The vertex K is the midpoint of side RS. What is the area of the shaded region?

问题 15

图中有两个正方形 $JKLM$ 和 $PQRS$。JK 的长是 8 厘米，PQ 的长是 6 厘米。顶点 K 是边 RS 的中点。问阴影部分的面积是多少？

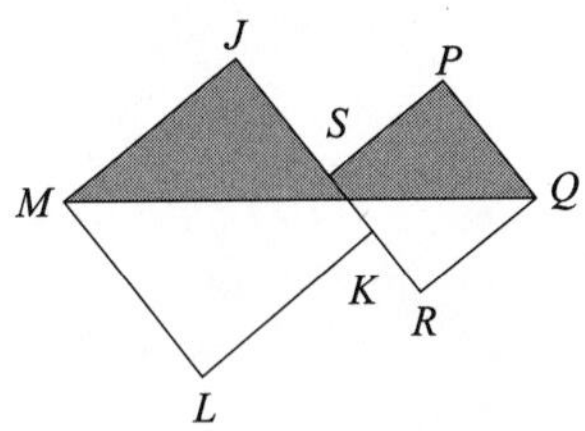

(A) 42cm^2 (B) 44cm^2 (C) 46cm^2 (D) 48cm^2 (E) 47cm^2

Problem 16

The areas of the two rectangles in the diagram are 25cm^2 and 13cm^2 as indicated. What is the value of x?

问题 16

如图所示，图中两个矩形的面积分别为 25cm^2 和 13cm^2。问 x 的值是多少？

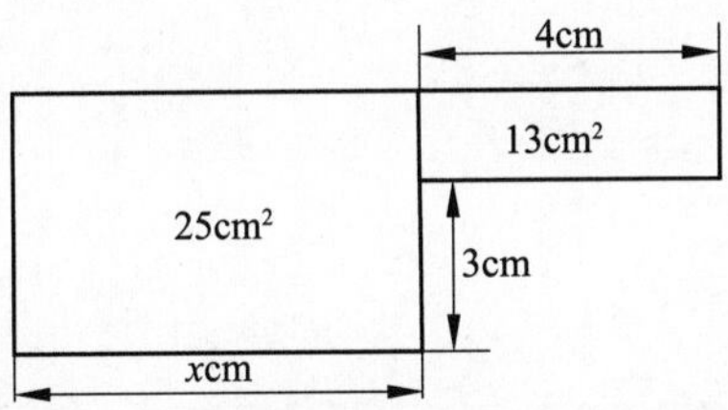

(A) 3 (B) 4 (C) 5 (D) 6 (E) 7

Problem 17

The shapes P and Q are formed from two and three identical rectangles, respectively. Their perimeters are 60 cm and 83 cm respectively. What is the perimeter of one of the rectangles?

问题 17

图形 P 和 Q 分别由两个和三个相同的矩形组成。它们的周长分别是 60 厘米和 83 厘米。问其中一个长方形的周长是多少？

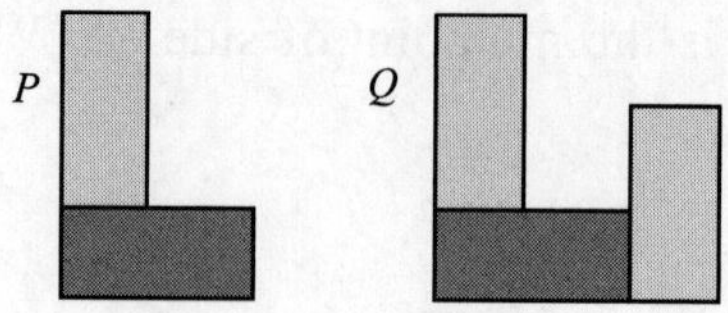

(A) 36cm　　(B) 37cm　　(C) 38cm　　(D) 39cm　　(E) 40cm

Chapter 10　Circles

第 10 章　圆

10.1 Terms 名词定义

Chord: a line segment joining two points on the circle.

Diameter: a chord through the center.

Radius: a line segment joining the center to a point on the circumference.

Tangent: a line that touches the circle at one and only one point.

Secant: a line that intersects the circle at two points.

Minor arc: arc that is less than a semicircle.

Major arc: arc that is greater than a semicircle.

Concentric circles: circles that have the same center.

弦： 连接圆上两点的线段。

直径： 穿过圆心的弦。

半径： 连接圆心和圆周上某一点的线段。

切线： 和圆只有一个交点的直线。

正割： 和圆相交于两点的直线。

小弧： 小于半圆的弧。

大弧： 大于半圆的弧。

同心圆： 圆心相同的圆。

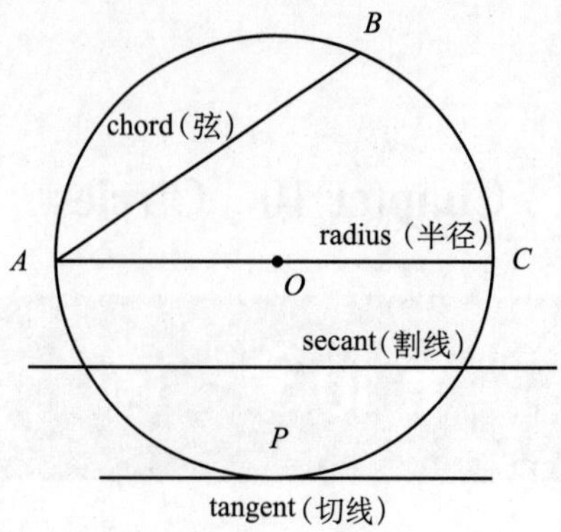

10.2 Formulas 公式

10.2.1 Diameter 直径

The diameter of a circle is twice its radius.

$$d = 2r$$

where "r" is the radius of the circle, "d" is the diameter.

圆的直径是半径的两倍。

$$d = 2r$$

其中"r"是圆的半径,"d"是直径。

10.2.2 Circumference of a Circle 圆的周长

$$C = 2\pi r$$

where "r" is the radius of the circle, "C" is the circumference, and π is the mathematical constant whose value is approximated to 3.14 or $\frac{22}{7}$.

$$C = 2\pi r$$

其中"r"是圆的半径,"C"是圆周长,π 是数学常数,其值近似为 3.14 或 $\frac{22}{7}$。

10.2.3 Area of a Circle 圆的面积

$$A = \pi r^2$$

where "r" is the radius of the circle, "A" is the area, and π is the mathematical constant whose value is approximated to 3.14 or $\frac{22}{7}$.

$$A = \pi r^2$$

其中“r”是圆的半径，“A”是圆的面积，π 是数学常数，其值近似为 3.14 或 $\dfrac{22}{7}$。

10.2.4 Area of a Sector 扇形的面积

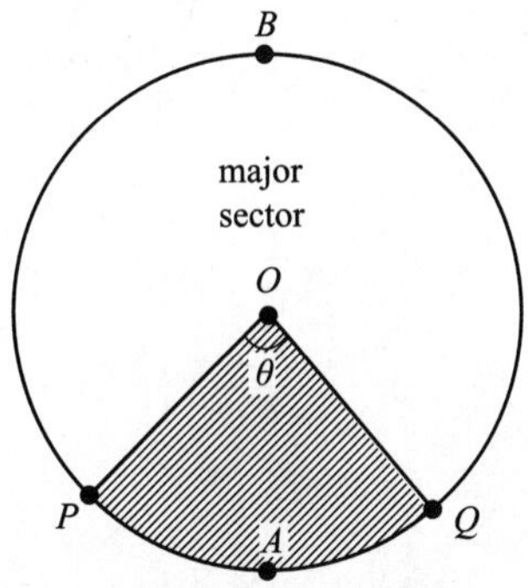

In a circle with radius r and center at O, let $\angle POQ = \theta$ (in degrees) be the angle of the sector.

When the angle of the sector is 360° (i.e., the whole circle), then the area of the sector is: $A = \pi r^2$.

When the angle is 1°, then the area of a sector is: $A = \dfrac{1\pi r^2}{360°}$.

So, when the angle is θ, area of sector $OPAQ$ is: $A = \dfrac{\theta}{360°} \times \pi r^2$.

在半径为 r 圆心位于 O 的圆上，设 $\angle POQ = \theta$（单位：度）为扇形角。

当扇形的角度为 360°（即整个圆）时，扇形的面积为 $A = \pi r^2$。

当角度为 1° 时，扇形的面积为 $A = \dfrac{1\pi r^2}{360°}$。

因此，当角度为 θ 时，扇形 $OPAQ$ 的面积为 $A = \dfrac{\theta}{360°} \times \pi r^2$。

10.2.5 Length of an Arc 扇形的弧长

Similarly, length of the arc (PQ) of the sector with angle θ is:

$$L = \frac{\theta}{360°} \times 2\pi r$$

同理，角度为 θ 的扇形的弧长（PQ）：

$$L = \frac{\theta}{360°} \times 2\pi r$$

Example 1

The figure below shows a large white circle with a number of smaller white and shaded circles in its interior. What fraction of the interior of the large white circle is shaded?

例题 1

下图显示了一个大白圆，其内部有一些小白圆和阴影圆。大白圆内部阴影部分的比例是多少？

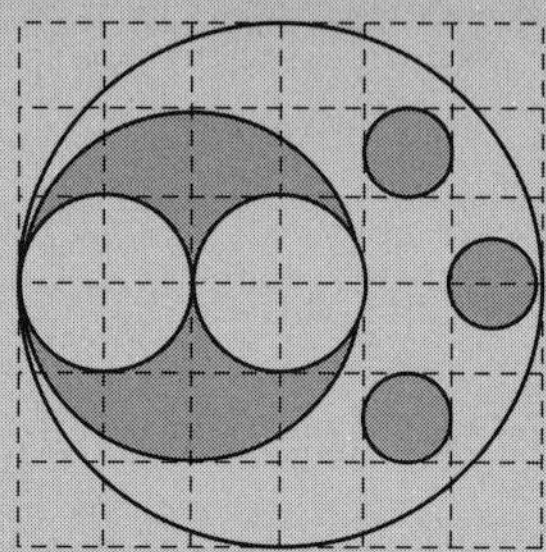

(A) $\frac{1}{4}$　(B) $\frac{11}{36}$　(C) $\frac{1}{3}$　(D) $\frac{19}{36}$　(E) $\frac{5}{9}$

Solution

According to the formula for the area of a circle, the total area of the radius 3 circle is 9π. Similarly, the area of a circle of radius 0.5 is: $A_1=\pi\left(\frac{1}{2}\right)^2=\frac{1}{4}\pi$, The area of a circle of radius 2 is: $A_2=\pi\cdot 2^2=4\pi$, and The area of a circle with radius 1 is: $A_3=\pi\cdot 1^2=\pi$. This can be done by adding the areas of three circles with radius 0.5 and the area of the circle with radius 2, then subtract the areas of the two circles with radius 1. So we get $3\times\frac{1}{4}\pi+4\pi-\pi-\pi=\frac{3}{4}\pi+2\pi=\frac{11\pi}{4}$. Therefore, the ratio is $\frac{\frac{11\pi}{4}}{9\pi}=$ (B) $\frac{11}{36}$.

解　析

根据圆的面积公式，最大的圆的半径为 3，则大圆面积为 9π。同理，半径为 0.5 的圆的面积为 $A_1=\pi\left(\frac{1}{2}\right)^2=\frac{1}{4}\pi$，半径为 2 的圆的面积为 $A_2=\pi\cdot 2^2=4\pi$，半径为 1 的

圆的面积为 $A_3=\pi\cdot 1^2=\pi$。根据图计算阴影部分面积可得，将三个半径为 0.5 的圆的面积和半径为 2 的圆的面积相加，然后减去两个半径为 1 的圆的面积，就可以得到阴影面积 $3\times\frac{1}{4}\pi+4\pi-\pi-\pi=\frac{3}{4}\pi+2\pi=\frac{11\pi}{4}$。因此，比值是 $\frac{\frac{11\pi}{4}}{9\pi}=\frac{11}{36}$，故选 (B)。

Example 2

Four pepperoni circles will exactly fit across the diameter of a 16-inch pizza when placed. If 8 circles of pepperoni are placed on this pizza without overlap, what fraction of the pizza is covered by pepperoni?

例题 2

在一个直径为 16 英寸的比萨上放上 4 个意大利辣香肠圈，直径正好相等。如果在比萨上一共放上 8 个意大利辣香肠圈且不重叠，那么意大利辣香肠圈一共占比萨面积的几分之几？

(A) $\frac{1}{2}$ (B) $\frac{2}{3}$ (C) $\frac{3}{4}$ (D) $\frac{5}{6}$ (E) $\frac{7}{8}$

Solution

A pepperoni circle's diameter is $\frac{16}{4}=4$. So the area of the 8 circles of pepperoni is $8\cdot\pi\left(\frac{4}{2}\right)^2=32\pi$. The pizza's area is $\pi\left(\frac{16}{2}\right)^2=64\pi$. Therefore, the ratio is $\frac{32\pi}{64\pi}=(\text{A})\frac{1}{2}$.

解 析

圆形香肠片的直径是 $\frac{16}{4}=4$。所以，8 个圆形香肠片的面积是 $8\cdot\pi\left(\frac{4}{2}\right)^2=32\pi$。比萨的面积是 $\pi\left(\frac{16}{2}\right)^2=64\pi$。因此，比例是 $\frac{32\pi}{64\pi}=\frac{1}{2}$，故选 (A)。

Example 3

Rectangle $ABCD$ is inscribed in a semicircle with diameter FE as shown in the figure. Let $DA=16$ and $FD=AE=9$. What is the area of $ABCD$?

例题 3

如图所示，矩形 $ABCD$ 嵌入直径为 FE 的半圆中。设 $DA = 16$，$FD = AE = 9$。$ABCD$ 的面积是多少？

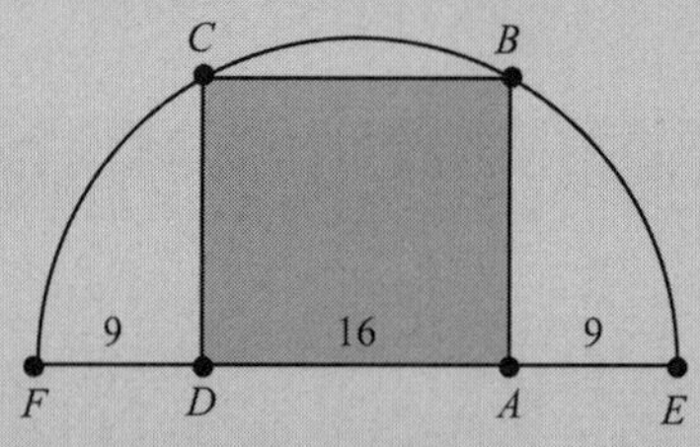

(A) 240　　(B) 248　　(C) 256　　(D) 264　　(E) 272

Solution

Let O be the center of the semicircle. The diameter of the semicircle is $9+16+9=34$, so $OC = 17$. By symmetry, O is the midpoint of DA, so $OD = OA = 16/2 = 8$. Applying Pythagorean theorem to the right-angled triangle $\triangle ODC$, we have that $CD=\sqrt{17^2-8^2}=15$. Thus, the area of $ABCD$ is $16 \times 15 =$ (A) 240.

解　析

设 O 为半圆的圆心。半圆的直径是 $9+16+9=34$，所以 $OC=17$。根据对称性，O 是 DA 的中点，所以 $OD=OA=\dfrac{16}{2}=8$。根据直角 $\triangle ODC$ 中的勾股定理，可得 $CD=\sqrt{17^2-8^2}=15$。因此，$ABCD$ 的面积为 $16 \times 15 = 240$，故选 (A).

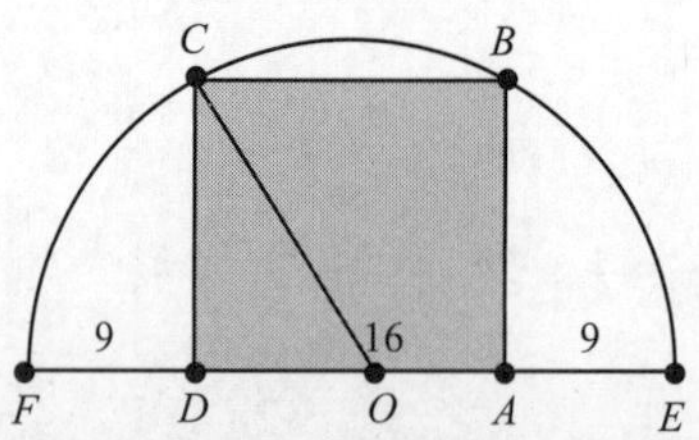

Example 4

The following figures are composed of squares and circles. Which figure has a shaded region with largest area?

例题 4

下列图形由正方形和圆形组成。哪个图形的阴影部分面积最大?

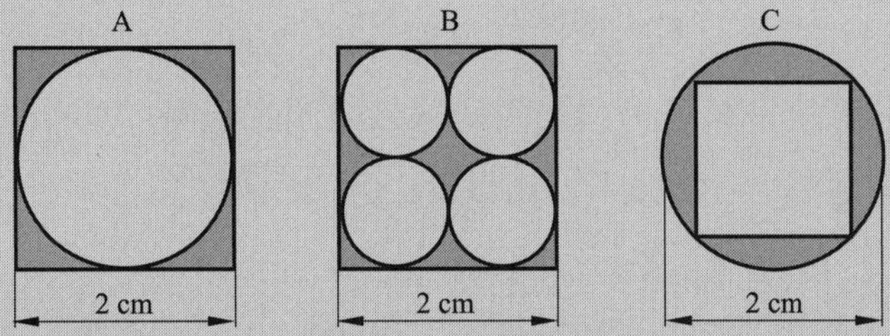

(A) A only (B) B only (C) C only

(D) both A and B (E) all are equal

Solution

A has shaded area of $2^2-1^2\pi=4-\pi\approx 0.86$.

B has shaded area of $2^2-4\times 0.5^2\pi=4-\pi\approx 0.86$.

C has shaded area of $1^2\pi-(2\times 2)/2=\pi-2\approx 1.14$, since the area of the square is $(d_1\times d_2)/2$, where d_1 and d_2 are the lengths of diagonals.

Therefore, the answer is (C).

解 析

A 的阴影面积是 $2^2-1^2\pi=4-\pi\approx 0.86$。

B 的阴影面积是 $2^2-4\times 0.5^2\pi=4-\pi\approx 0.86$。

C 的阴影面积是 $1^2\pi-(2\times 2)/2=\pi-2\approx 1.14$,正方形的面积公式是 $(d_1\times d_2)/2$,其中 d_1 和 d_2 是对角线的长度。

因此,答案是 (C)。

Example 5

Three concentric circles centered at O have radii of 1, 2 and 3. Points B and C lie on the largest circle. The region between the two smaller circles is shaded, as is the portion of the region between the two larger circles bounded by central angles BOC, as shown in the figure below. Suppose the ratio of shaded area to unshaded area is 5 : 4.

What is the measure of $\angle BOC$ in degrees?

例题 5

以 O 为圆心的三个同心圆的半径分别为 1，2 和 3。点 B 和 点 C 位于最大的圆上。如下图所示，两个小圆之间的区域是阴影区域，两个大圆之间以圆心角 BOC 为界的部分区域也是阴影区域。假设阴影区域和非阴影区域的面积比是 5∶4，那么 $\angle BOC$ 度数是多少？

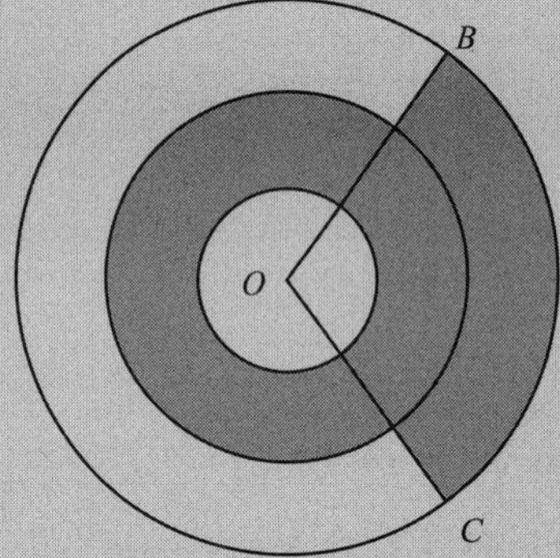

(A) 108　　(B) 120　　(C) 135　　(D) 144　　(E) 150

Solution

Let $\angle BOC = x$. We see that the shaded region is composed of a ring and part of a sector subtending $\angle BOC$. The area of the ring is $4\pi - \pi = 3\pi$ and the area of the shaded part of the sector is $\frac{x}{360}\cdot(9\pi - 4\pi) = \frac{x}{72}\pi$, so the area of the shaded region is $\left(3+\frac{x}{72}\right)\pi$. The area of the unshaded region is $9\pi - \left(3+\frac{x}{72}\right)\pi = \left(6-\frac{x}{72}\right)\pi$. Given the ratio of area is 5∶4, we have $4\cdot\left(3+\frac{x}{72}\right)\pi = 5\cdot\left(6-\frac{x}{72}\right)\pi$. Solving for x we get $x =$ (D) 144.

解　析

设 $\angle BOC = x$，由图可得，阴影区域是由一个环和扇形区域的一部分组成，该扇形区域对着 $\angle BOC$。环的面积是 $4\pi - \pi = 3\pi$，扇形的阴影部分面积为 $\frac{x}{360}\cdot(9\pi - 4\pi) = \frac{x}{72}\pi$，因此阴影的面积是 $\left(3+\frac{x}{72}\right)\pi$。由此可得，非阴影部分的面积

为 $9\pi-\left(3+\frac{x}{72}\right)\pi=\left(6-\frac{x}{72}\right)\pi$。根据面积比是 5 : 4 可得 $4\cdot\left(3+\frac{x}{72}\right)\pi=5\cdot\left(6-\frac{x}{72}\right)\pi$，解得 $x=144$，故选 (D)。

Example 6

A circle with radius 10 is inscribed in a square and circumscribed about another square as shown. Which fraction is closest to the ratio of the circle's shaded area to the area between the two squares?

例题 6

如图所示，一个半径为 10 的圆被刻在一个正方形内，同时一个小白正方形以圆的直径为对角线内接在圆里。哪个分数最接近圆中的阴影面积与两个正方形之间的面积之比？

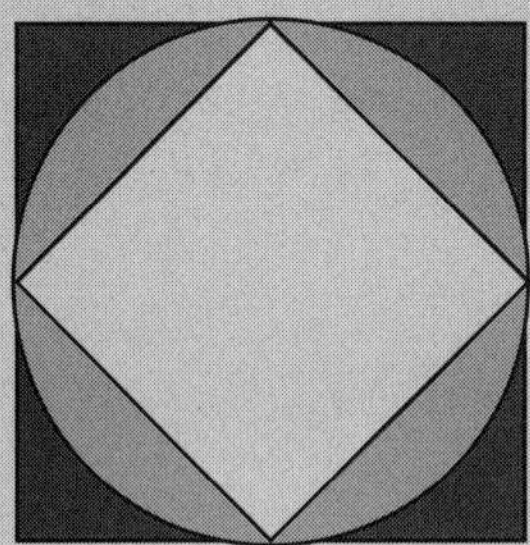

(A) $\frac{1}{2}$ (B) 1 (C) $\frac{3}{2}$ (D) 2 (E) $\frac{5}{2}$

Solution

The circle's shaded area equals the difference of the areas of the samller square and the circle, which is $100\pi-20\times 20\times 0.5=100\pi-200$, where 20 is the length of the diagonal of the smaller square.

If you draw the diagonals of the smaller square, you will see that the larger square is split into 4 congruent half-shaded squares. So, the area between the squares equals to the area of the smaller square: 200.

Approximating π to 3.14, we can get the ratio of the circle's shaded area to the area between the two squares is about $\frac{100\pi-200}{200}\approx\frac{314-200}{200}=\frac{114}{200}\approx$ (A) $\frac{1}{2}$.

解 析

圆中的阴影面积等于圆的面积减去小正方形的面积，即 $100\pi-20\times20\times0.5=100\pi-200$，其中 20 是小正方形对角线的长度。

如果画出小正方形的对角线，就会发现大正方形被分割成 4 个相等的半阴影正方形。所以两个正方形之间的面积等于小正方形的面积：200。

将π近似为 3.14，可得圆的阴影部分的面积与两个正方形之间的面积之比约为 $\frac{100\pi-200}{200}\approx\frac{314-200}{200}=\frac{114}{200}\approx\frac{1}{2}$，故选 (A)。

Example 7

A roll of tape is 4 inches in diameter and is wrapped around a ring that is 2 inches in diameter. A cross section of the tape is shown in the figure below. The tape is 0.02 inches thick. If the tape is completely unrolled, approximately how long would it be? Round your answer to the nearest 100 inches.

例题 7

一卷直径为 4 英寸的胶带缠绕在一个直径为 2 英寸的圆环上。胶带的横截面如下图所示。胶带的厚度为 0.02 英寸。如果胶带完全展开，大约有多长？将结果四舍五入到最接近的 100 英寸。

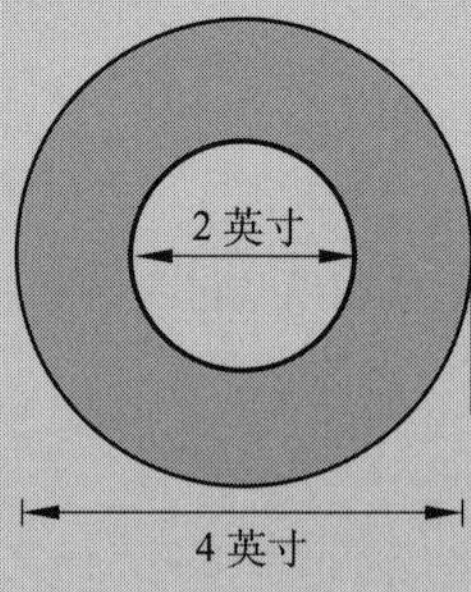

(A) 400　(B) 500　(C) 600　(D) 1000　(E) 1200

Solution

The roll of tape is 1/0.02 = 50 layers thick. To find the total length, we have to find the average circumference of each concentric circle and multiply it by 50. Since the diameter of the small circle is 2 inches and the diameter of the large one is 4 inches, the "middle value" is 3. Therefore, the average circumference is 3π, and the length is $3\pi \cdot 50 = 150\pi \approx$ (B) 500.

解 析

这卷胶带的厚度是 1/0.02 = 50 层。为了求出总长度，必须求出每个同心圆的平均周长，然后乘以 50。已知小圆的直径是 2 英寸，大圆的直径是 4 英寸，可取平均数 3 作为直径。因此，平均周长是 3π。所以 $3\pi \cdot 50 = 150\pi \approx 500$，故选 (B)。

Example 8

Mary's winning art design is shown. The smallest circle has radius 3 inches, with each successive circle's radius increasing by 1 inch. Approximately what percent of the design is black?

例题 8

玛丽的获奖作品如图所示。最小的圆半径为 3 英寸，以后每一个圆的半径都增加 1 英寸。黑色部分的面积大约占整个面积的百分之多少？

(A) 43%　　(B) 44%　　(C) 46%　　(D) 47%　　(E) 48%

Solution

Circle #	Radius	Area
1	3	9π
2	4	16π
3	5	25π
4	6	36π
5	7	49π
6	8	64π

As shown in the table, the entire circle's area is 64π. The area of the black regions is $9\pi+(25-16)\pi+(49-36)\pi=31\pi$. Thus, the percentage of the black region is $\frac{31\pi}{64\pi}\approx$ (E)48.

解　析

圆编号	半径	面积
1	3	9π
2	4	16π
3	5	25π
4	6	36π
5	7	49π
6	8	64π

如表所示，最大圆的面积是64π。黑色区域的面积是$9\pi+(25-16)\pi+(49-36)\pi=31\pi$。所以黑色区域所占的百分比是$\frac{31\pi}{64\pi}\approx 48$，故选(E)。

Example 9

A ball with diameter 2 inches starts at point A to roll along the track shown. The track is comprised of 3 semicircular arcs whose radii are $R_1=90$ inches, $R_2=40$ inches, and $R_3=60$ inches, respectively. The ball always remains in contact with the track and does not slip. What is the distance the center of the ball travels over the course from A to B?

例题 9

一个直径为 2 英寸的小球从 A 点开始沿着如图所示的轨道滚动。轨道由 3 个半圆弧组成，半径分别为 $R_1 = 90$ 英寸、$R_2 = 40$ 英寸和 $R_3 = 60$ 英寸。小球始终与轨道接触不会滑动。球从 A 到 B 滚动时中心移动的距离是多少?

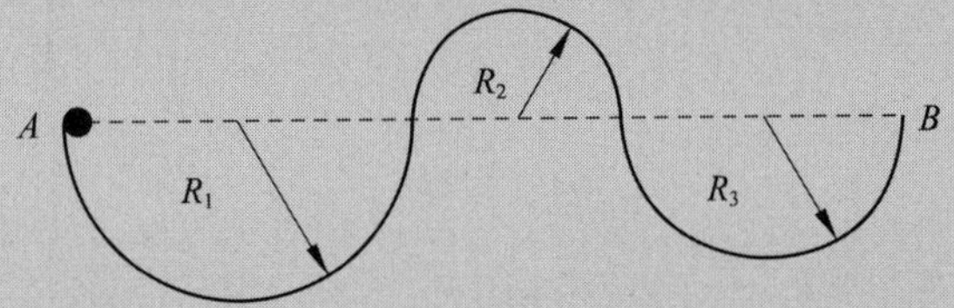

(A) 189π　　(B) 190π　　(C) 200π　　(D) 220π　　(E) 233π

Solution

Notice that the question is asking for the distance the "center" traveled, not the length of the track. The ball has a radius of 1, so the radii of the trace should be 1 more or 1 less than the radii of the semicircle tracks. The total length of all arcs traveled by the center is $(90-1)\pi+(40+1)\pi+(60-1)\pi=$ (A) 189π.

解　析

题目问的是"球心"移动的距离，而不是轨道的长度。小球的半径为 1，那么球心所走路径的半径应该在原基础上减 1 和加 1，因此球心轨迹的所有弧线的总长度为 $(90-1)\pi+(40+1)\pi+(60-1)\pi=189\pi$，故选 (A)。

Example 10

In the figure shown, US and UT are line segments each of length 3, and $\angle TUS$=60°. Arcs TR and SR are each one-sixth of a circle with radius 3. What is the area of the region shown?

例题 10

如图所示，US 和 UT 分别是长度为 3 的线段，$\angle TUS$=60°。弧 TR 和 SR 分别是半径为 3 的圆的 $\frac{1}{6}$。图中区域的面积是多少?

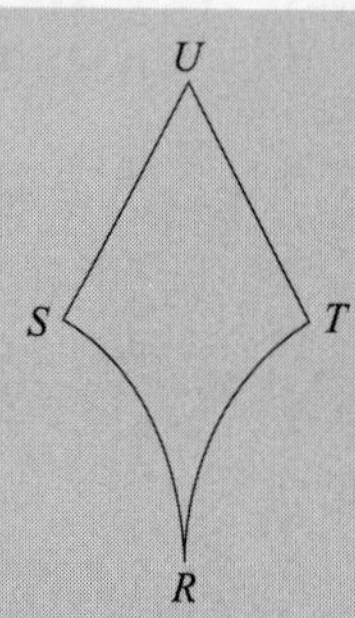

(A) $\sqrt{3}-3\pi$ (B) $9\sqrt{3}-3\pi$ (C) $3\sqrt{3}-\frac{1}{3}\pi$

(D) $9\sqrt{3}+\pi$ (E) $9\sqrt{3}-\frac{2}{3}\pi$

Solution

Extend *US* and *UT* to form an equilateral triangle with side lenth 3 + 3 = 6. Then the area we are looking for equals the difference between the area of the triangle and areas of the two 60-degree sectors. Recall that the formula for the area of an equilateral triangle is $\frac{\sqrt{3}}{4}a^2$, where a is the side length. Therefore, the area is $\frac{\sqrt{3}}{4}\times 6^2-2\times\frac{60}{360}\times\pi\times 3^2=9\sqrt{3}-3\pi$. Therefore, the answer is (B).

解 析

延长 *US* 和 *UT* 构建边长为 3 + 3 = 6 的等边三角形。所求面积等于大等边三角形的面积减去两个 60° 扇形的面积。等边三角形的面积公式 $\frac{\sqrt{3}}{4}a^2$，其中 a 是边长。因此，所求面积为 $\frac{\sqrt{3}}{4}\times 6^2-2\times\frac{60}{360}\times\pi\times 3^2=9\sqrt{3}-3\pi$，故选 (B)。

PRACTICE 练习

Problem 1

A square and a circle have the same area. What is the ratio of the side length of the square to the radius of the circle?

问题 1

一个正方形和一个圆的面积相同。那么正方形的边长与圆的半径的比是多少？

(A) $\frac{\sqrt{2}}{2}$ (B) $\sqrt{\pi}$ (C) π (D) 2π (E) π^2

Problem 2

In the diagram below, the diameter of each of the two smaller circles equals the radius of the larger circle. If the two smaller circles have a combined area of 2 square units, what is the area of the shaded region, in square units?

问题 2

在下图中，两个小圆的直径等于大圆的半径。如果两个小圆的面积之和是 2 平方单位，那么阴影部分的面积是多少平方单位？

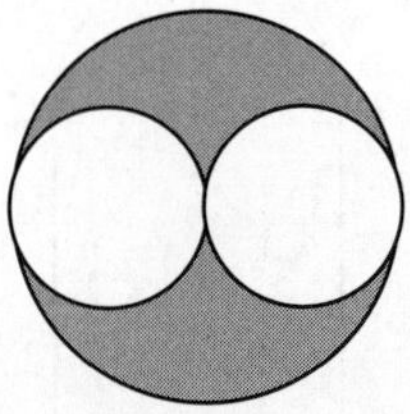

(A) $\frac{1}{2}$ (B) $\frac{3}{2}$ (C) $\frac{3}{2}$ (D) 2 (E) 1

Problem 3

The circumference of the circle with center O is divided into 12 equal arcs, marked by letters A through L as seen below. What is the number of degrees in the difference of the angles x and y?

问题 3

如下图所示，以 O 为圆心的圆的周长被分成 12 条相等的弧，并标有字母 $A \sim L$。角 x 和角 y 的度数的差是多少？

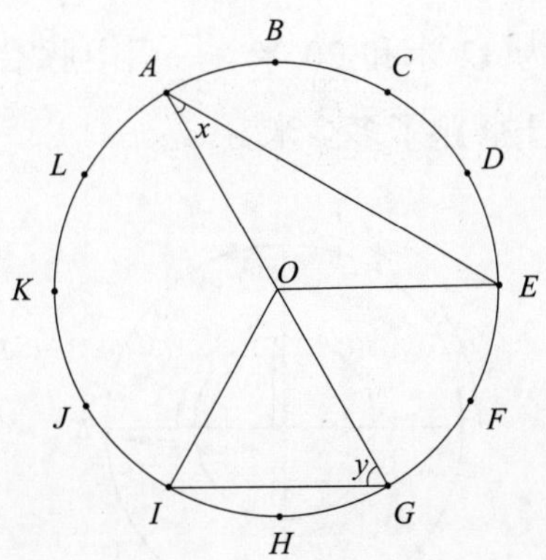

(A) 10 (B) 15 (C) 30 (D) 60 (E) 90

Problem 4

A decorative window is made up of a rectangle with semicircles on either end. The ratio of AD to AB is $5:4$, and AB is 40 inches. What is the ratio of the area of the rectangle to the combined areas of the semicircles?

问题 4

一个装饰窗由一个矩形和两端的半圆形组成。AD 与 AB 之比为 $5:4$，AB 的长为 40 英寸。矩形面积与两个半圆形面积之和的比是多少？

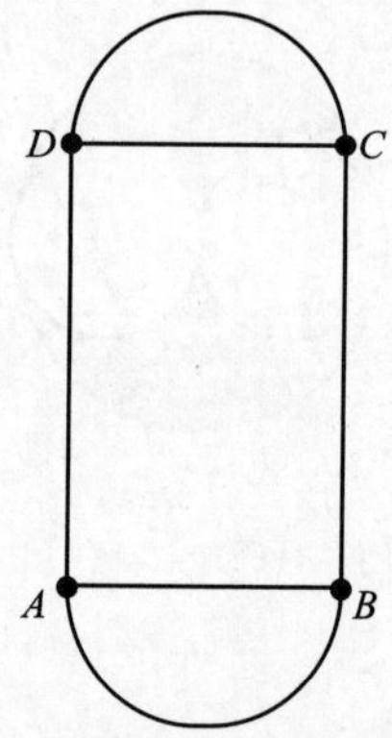

(A) $20:\pi$ (B) $5:4$ (C) $5:\pi$ (D) $5:4\pi$ (E) $3:2\pi$

Problem 5

Two circles that share the same center have radii 12 meters and 20 meters. An aardvark runs along the path shown, starting at A and ending at K. How many meters does the aardvark run?

问题 5

两个同心圆的半径分别为 12 米和 20 米。一只土豚沿着如图所示的路径奔跑，从 A 处开始，到 K 处结束。问土豚跑了多少米？

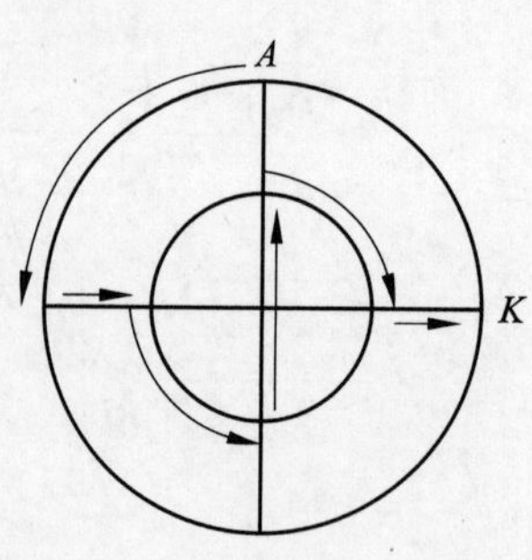

(A) $12\pi+20$ (B) $20\pi+30$ (C) $20\pi+40$ (D) $22\pi+20$ (E) $22\pi+40$

Problem 6

Rectangle $ABCD$ has sides $CD=3$ and $DA=5$. A circle of radius 1 is centered at A, a circle of radius 2 is centered at B, and a circle of radius 3 is centered at C. Which of the following is closest to the area of the region inside the rectangle but outside all three circles?

问题 6

矩形 $ABCD$ 的边长分别为 $CD=3$ 和 $DA=5$。半径为 1 的圆以 A 为圆心，半径为 2 的圆以 B 为圆心，半径为 3 的圆以 C 为圆心。以下哪个数值最接近矩形内部不被圆覆盖到的区域的面积？

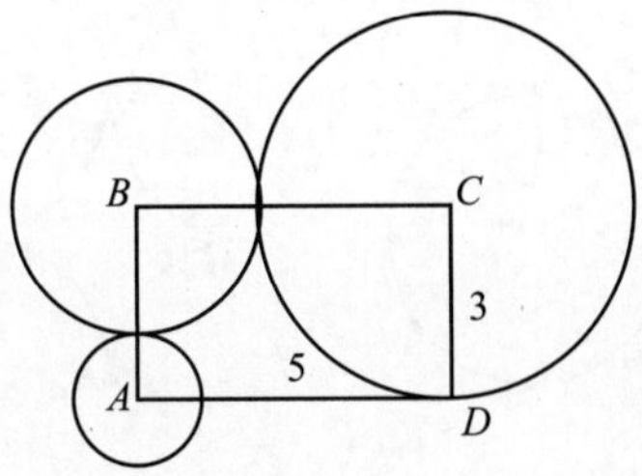

(A) 3.5 (B) 4.0 (C) 4.5 (D) 5.0 (E) 5.5

Problem 7

A 1×2 rectangle is inscribed in a semicircle with longer side on the diameter. What is the area of the semicircle?

问题 7

一个 1×2 的矩形嵌在一个半圆里，矩形的长边在圆的直径上。问半圆的面积是多少？

(A) $\frac{5}{2}\pi$ (B) $\frac{25}{4}\pi$ (C) $\frac{25}{2}\pi$ (D) 5π (E) 25π

Problem 8

Two congruent circles centered at points A and B each pass through the other circle's center. The line containing both A and B is extended to intersect the circles at points C and D. The circles intersect at two points, one of which is E. What is the degree measure of $\angle ECD$?

问题 8

以点 A 和点 B 为圆心的两个全等圆分别经过另一个圆的圆心。将包含点 A 和点 B

的直线延长，与圆相交于点 C 和点 D。两圆相交于两点，其中一点为 E。问$\angle ECD$ 的度数是多少?

(A) 30　　(B) 40　　(C) 50　　(D) 60　　(E) 90

Problem 9

A circle of radius 4 is cut into four congruent arcs. The four arcs are joined to form the star figure shown. What is the ratio of the area of the star figure to the area of the original circle?

问题 9

半径为 4 的圆被切割成四条全等的弧。将四条弧连接起来形成如图所示的星形图形。星形图形的面积与原来的圆的面积的比是多少?

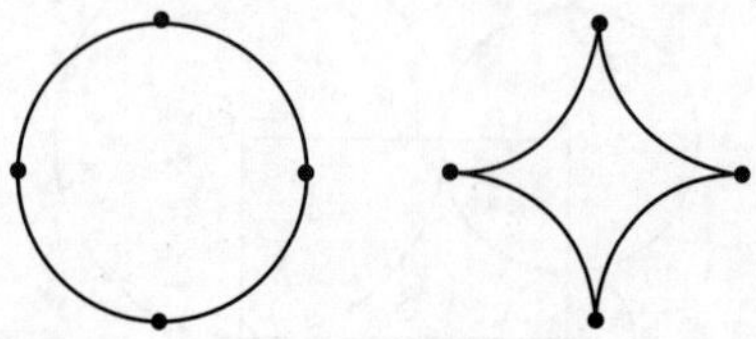

(A) $\frac{4-\pi}{\pi}$　　(B) $\frac{1}{\pi}$　　(C) $\frac{\sqrt{2}}{\pi}$　　(D) $\frac{\pi-1}{\pi}$　　(E) $\frac{3}{\pi}$

Problem 10

$\angle ABC$ of $\triangle ABC$ is a right angle. The sides of $\triangle ABC$ are the diameters of semicircles as shown. The area of the semicircle on AB equals 4.5π, and the arc of the semicircle on AC has length 5π. What is the radius of the semicircle on BC?

问题 10

在$\triangle ABC$ 中，$\angle ABC$ 是直角。如图所示，$\triangle ABC$ 的边是半圆的直径。以 AB 为直径的半圆的面积是 4.5π，以 AC 为直径的半圆的弧长为 5π。问以 BC 为直径的半圆的半径是多少?

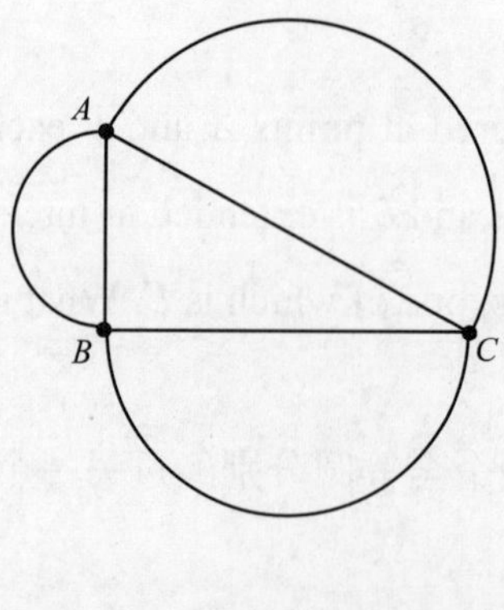

(A) 3.5 (B) 4 (C) 5 (D) 8 (E) 9

Problem 11

Circle X has a radius of 2π. Circle Y has a circumference of 6π. Circle Z has an area of 8π. List the circles in order from smallest to largest radius.

问题 11

圆 X 的半径为 2π。圆 Y 的周长为 6π。圆 Z 的面积是 8π。按半径从小到大的顺序将这 3 个圆排序。

(A) XYZ (B) ZYX (C) YXZ (D) ZXY (E) XZY

Problem 12

Two 8×8 squares intersect at right angles, bisecting their intersecting sides, as shown. The circle's diameter is the segment between the two points of intersection. What is the area of the shaded region created by removing the circle from the squares?

问题 12

如图所示，两个 8×8 的正方形以直角相交，将相交的边互相平分。圆的直径是两个交点之间的连线。将圆从正方形中移除后，阴影部分的面积是多少？

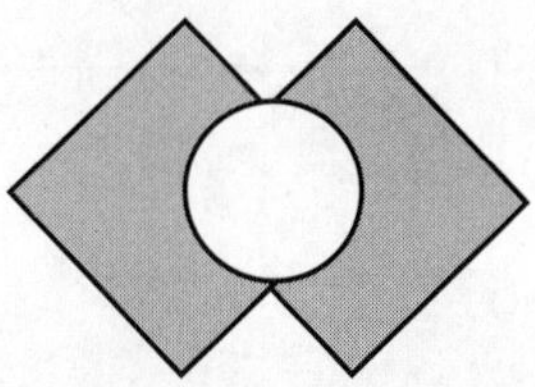

(A) $64-8\pi$ (B) $56-8\pi$ (C) $64-4\pi$

(D) $112-8\pi$ (E) $112-4\pi$

Problem 13

Sergei skated around an ice rink, gliding along different paths. The gray lines in the figures below show four of the paths labeled P, Q, R, and S. What is the sorted order of the four paths from longest to shortest?

问题 13

谢尔盖在溜冰场上滑冰，沿着不同的路径滑行。下图中的灰线表示其中标有 P、Q、R 和 S 的四条路径。这四条路径长到短的顺序是什么？

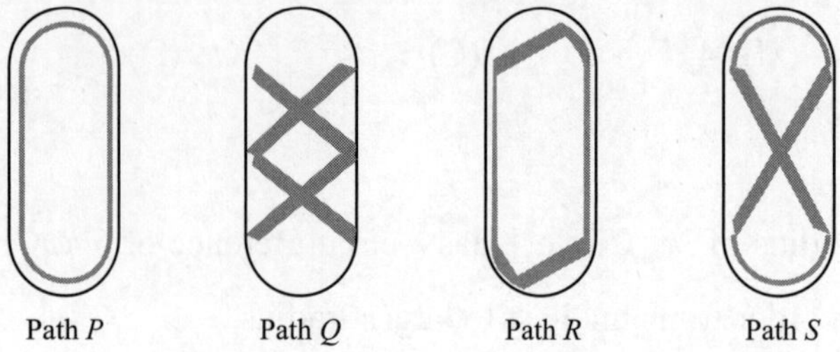

(A) $SQRP$ (B) $QSRP$ (C) $SQPR$

(D) $QSPR$ (E) They're all the same length 一样长

Problem 14

Semicircles POQ and ROS pass through the center O. What is the ratio of the combined areas of the two semicircles to the area of circle O?

问题 14

半圆 POQ 和 ROS 经过圆心 O。两个半圆的面积总和与圆 O 的面积之比是?

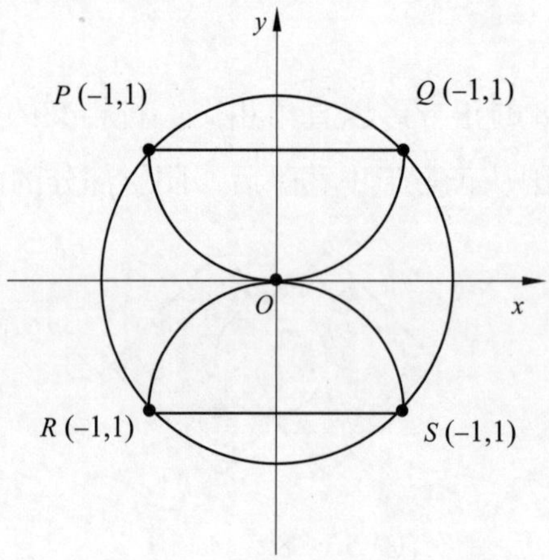

(A) $\frac{\sqrt{2}}{4}$ (B) $\frac{1}{2}$ (C) $\frac{2}{\pi}$ (D) $\frac{2}{3}$ (E) $\frac{\sqrt{2}}{2}$

Problem 15

A circle with radius $\frac{40}{\sqrt{\pi}}$ and a square share the same center. The total area of the regions that are inside the circle and outside the square is equal to the total area of the regions that are outside the circle and inside the square. What is the length of the side of the square?

问题 15

一个半径为 $\dfrac{40}{\sqrt{\pi}}$ 的圆和一个正方形共享同一个中心。在圆内但在正方形外的面积等于在圆外但在正方形内的面积。问正方形的边长是多少？

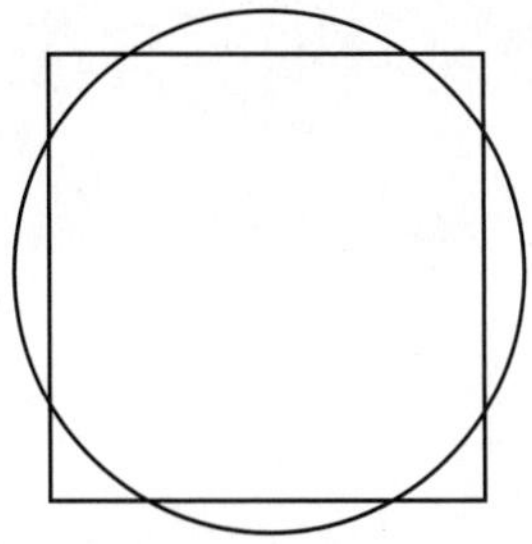

(A) 40　　(B) 32　　(C) 4π　　(D) 10　　(E) 40π

Chapter 11　Solid Geometry

第 11 章　立体几何

11.1　Basic Knowledge 基础内容

11.1.1　Rectangular Solid 长方体

Volume (V) 体积

$$V = abc$$

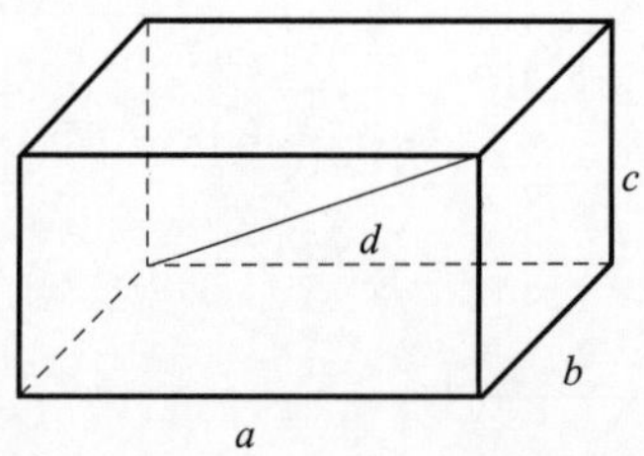

a: length of the base of the solid.

b: width of the base of the solid.

c: height of the solid.

d: space diagonal of the solid ($d^2 = a^2 + b^2 + c^2$).

a：长方体底部的长。

b：长方体底部的宽。

c：长方体的高。

d：长方体的空间对角线（ $d^2 = a^2 + b^2 + c^2$ ）。

Surface area (S) 表面积

$$S = 2(ab + bc + ac)$$

Note: For a cube, $a = b = c$.

注意：对于正方体，$a = b = c$。

11.1.2　Triangular Prism 三棱柱

Volume (V) 体积

$$V = \text{base area} \times \text{height} = \left(\frac{b \cdot h}{2}\right) \cdot l$$

b: base of the triangular base.

h: height of the triangular base.

l: height of the solid.

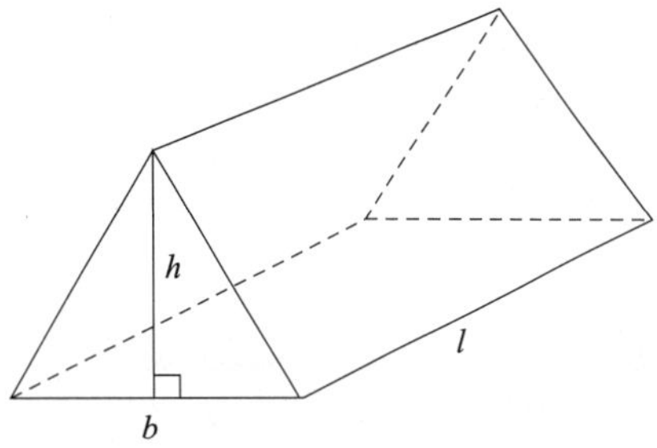

$$V = \text{底面积} \times \text{高} = \left(\frac{b \cdot h}{2}\right) \cdot l$$

b：三角形底座的底边。

h：三角形底座的高。

l：三棱柱的高。

Surface area (S) 表面积

S = area of two base triangles + area of three side rectangles.

$S = 2 \cdot \left(\frac{b \cdot h}{2}\right) + a \cdot l + b \cdot l + c \cdot l$, where a, b, and c are three side lengths of the base triangle.

S = 两个底面三角的面积 + 三个侧面矩形的面积。

$S = 2 \cdot \left(\frac{b \cdot h}{2}\right) + a \cdot l + b \cdot l + c \cdot l$，其中 a，b，c 分别是底面三角形的三条边长。

11.1.3 Cylinder 圆柱

Volume (V) 体积

$$V = \pi R^2 h = \frac{1}{4}\pi D^2 h$$

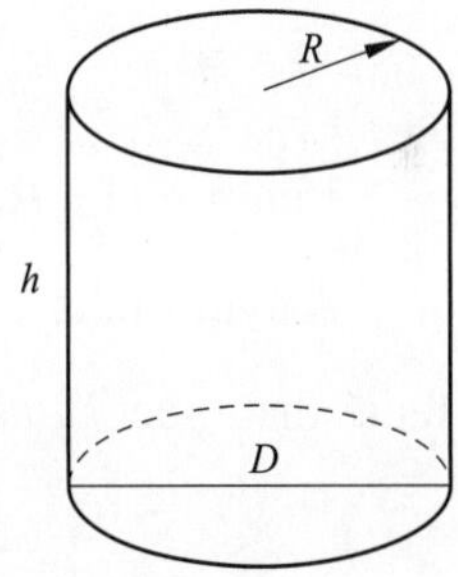

R: the radius of the circular base of the cylinder.

D: the diameter of the circular base of the cylinder.

h: the height of the cylinder.

R：底面圆的半径。

D：底面圆的直径。

h：圆柱的高。

Surface area (S) 表面积

$$S = 2\pi R^2 + 2\pi Rh = 2\pi R(R + h)$$

11.1.4 Cone 圆锥

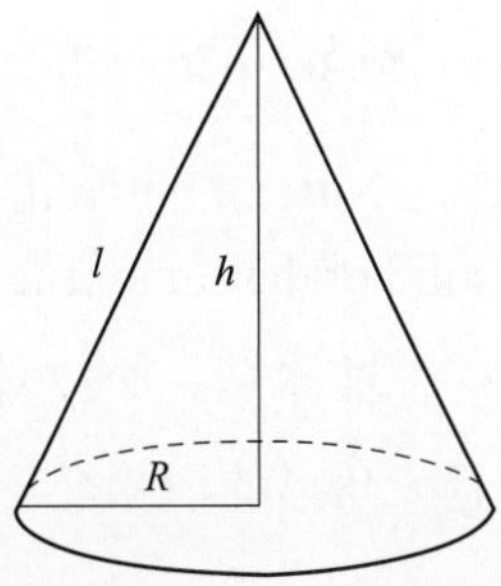

Volume (V) 体积

$$V = \frac{1}{3}\pi R^2 h = \frac{1}{12}\pi D^2 h$$

R: the radius of the circular base of the cylinder.

D: the diameter of the circular base of the cylinder.

h: the height of the cylinder.

l: the slant height of a right cone.

$$l = \sqrt{R^2 + h^2}$$

Note: The volume of the cone is $\frac{1}{3}$ of the volume of the cylinder of the same radius and height.

R：底面圆的半径。

D：底面圆的直径。

h：圆锥的高。

l：圆锥的斜高。

$$l = \sqrt{R^2 + h^2}$$

注意：圆锥的体积是和它拥有相同半径和高度的圆柱体积的 $\frac{1}{3}$。

Surface (S) 表面积

$$S = \pi Rl + \pi R^2$$

11.1.5 Pyramid 棱锥

$$V = \frac{1}{3}\text{base area} \times \text{height}$$

Volume of a Rectangular Pyramid (V) 底面为矩形的四棱锥体积

A pyramid is a solid figure with a polygonal base (in our case a rectangle) and triangular faces that meet at a common point (the apex).

$$V = \frac{1}{3}abc$$

a: the length of the base of the pyramid.

b: the width of the base of the pyramid.

c: the perpendicular height of the pyramid.

Note: The volume of the rectangular pyramid is $\frac{1}{3}$ of the volume of the rectangular solid of the same length, width, and height.

锥体是一个立体图形，它的底面是多边形（在例子中是矩形），侧面是三角形，且三角形的面相交于一个共同点（顶点）。

$$V = \frac{1}{3}abc$$

a：底面的长。

b：底面的宽。

c：垂直底面的高。

注意：底面为矩形的四棱锥体积是与它拥有相同长宽高的长方体的体积的 $\frac{1}{3}$。

Volume of a Regular Triangular Pyramid (V) 正三棱锥体积

$$V=\frac{1}{12}\sqrt{2}a^3$$

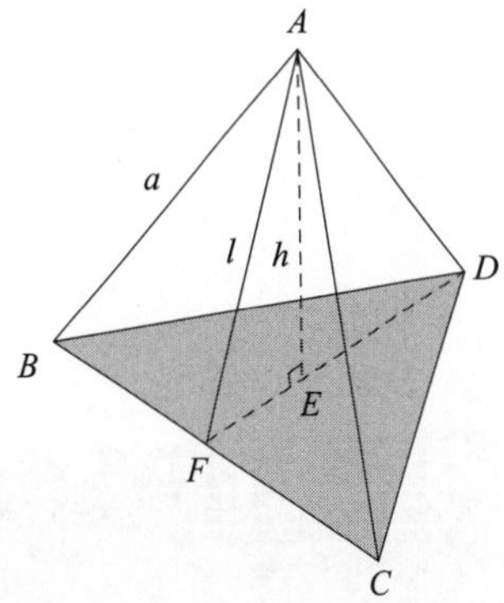

a: the edge length.

l: the slant height (the height of the equilateral triangle ABC)

$$l=\frac{1}{2}a\sqrt{3}$$

h: the height of the regular tetrahedron

$$h=\frac{1}{3}a\sqrt{6}$$

a：棱长

l：侧面的高（等边△ABC 的高）

$$l=\frac{1}{2}a\sqrt{3}$$

h：正三棱锥的高

$$h=\frac{1}{3}a\sqrt{6}$$

The surface area of the regular tetrahedron is simply four times the area of a single equilateral triangle face: $S=\sqrt{3}a^2$.

正四面体的表面积是四个等边三角形面积之和：$S=\sqrt{3}a^2$。

11.1.6 Octahedron 八面体

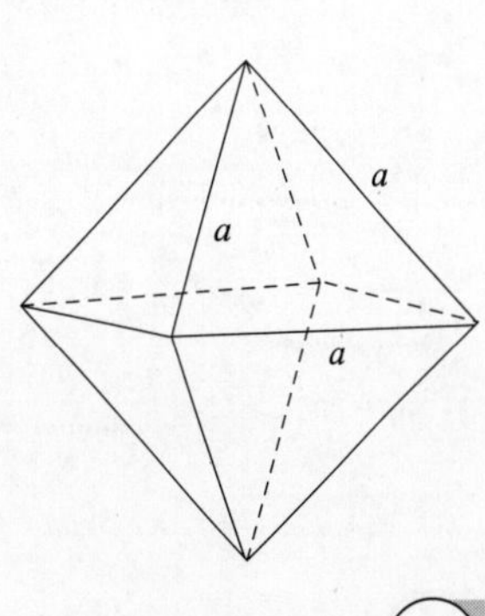

An Octahedron is a polyhedron having eight faces. Now let's take a look at the volume and the surface area of a regular octahedron as shown, where all sides are the same length (a).

八面体是有八个面组成的多面体。现在让我们来看看如图所示的正八面体的体积和表面积，其中所有边的长度都相同(a)。

Volume of Regular Octahedron (V) 正八面体体积

The volume is 2 times the volume of a square-base pyramid.

体积是正方形底四棱锥体积的 2 倍。

$$V=2\left(\frac{1}{3}a^2\cdot\frac{\sqrt{2}}{2}a\right)=\frac{\sqrt{2}}{3}a^3$$

Surface area of Regular Octahedron (S) 正八面体表面积

The surface area is 8 times the area of one face (an equilateral triangle).

表面积是一个面(等边三角形)面积的 8 倍。

$$S=8\cdot\frac{\sqrt{3}}{4}a^2=2\sqrt{3}a^2$$

Example 1

Akash's birthday cake is in the form of a $4\text{inch}\times4\text{inch}\times4\text{inch}$ cube. The cake has icing on the top and the four side faces, and no icing on the bottom. Suppose the cake is cut into 64 smaller cubes, each measuring $1\text{inch}\times1\text{inch}\times1\text{inch}$. How many small pieces will have icing on exactly one side?

例题 1

阿卡什的生日蛋糕是一个 4 英寸 ×4 英寸 ×4 英寸的立方体。蛋糕的顶部和 4 个侧面都有糖霜,底部没有糖霜。假设蛋糕被切成更小的立方体,每个立方体的大小为 1 英寸 ×1 英寸 ×1 英寸。问有多少小块蛋糕只有一面有糖霜?

(A) 16 (B) 22 (C) 36 (D) 28 (E) 32

Solution

Start by counting the 4 small cubes in the very center of the top layer that have icing on one side. Then look at the 4 small cubes in the center on all four sides and the two middle-bottom that have icing on one side, there are $4\times(4+2)=24$. So the total number is $4+24=$ (D) 28.

解 析

先数最上面一层的最中间 4 个小正方块有一面糖霜。再看 4 个侧面都是中间的 4 个小正方块和最下面一行中间两个是有一面有糖霜的，有 $4\times(4+2)=24$ 个。所以总数是 4 + 24 = 28，故选 (D)。

Example 2

How many pairs of parallel edges, such as *AB* and *GH* or *EH* and *FG*, does a cube have?

例题 2

一个正方体有多少对平行边，如 *AB* 和 *GH* 或 *EH* 和 *FG*？

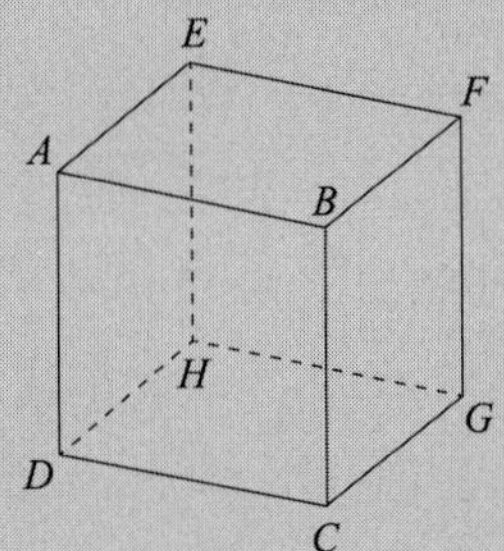

(A) 6 (B) 12 (C) 18 (D) 24 (E) 36

Solution

We first count the number of pairs of parallel lines that are in the same direction as *AB*. The pairs of parallel lines are *AB* and *EF*, *CD* and *GH*, *AB* and *CD*, *EF* and *GH*, *AB* and *GH*, and *CD* and *EF*. These are 6 pairs total. We can do the same for the lines in the same direction as *AE* and *AD*. This means there are $6\times3=18$ total pairs of parallel lines. Therefore, the answer is (C).

解 析

首先计算与 AB 方向相同的平行线对的数量。这几对平行线分别是 AB 和 EF，CD 和 GH，AB 和 CD，EF 和 GH，AB 和 GH 以及 CD 和 EF。一共是 6 对。对于与 AE 和 AD 方向相同的直线，可同理计算。因此，共有 $6\times3=18$ 对平行线，故选 (C)。

Example 3

Bill and David each have cats as pets. Bill buys cat food in cylindrical cans that are 8 cm in diameter and 15 cm high. David buys cat food in cylindrical cans that are 10 cm in diameter and 8 cm high. What is the ratio of the volume of one of Bill's cans to the volume of one of David cans?

例题 3

比尔和大卫都养了猫作为宠物。比尔买的猫粮是直径 8 厘米、高 15 厘米的圆柱形罐头。大卫买的猫粮是直径 10 厘米、高 8 厘米的圆柱形罐头。比尔买的罐子的体积与大卫的体积之比是多少?

(A) 6 : 5　　(B) 5 : 6　　(C) 3 : 2　　(D) 2 : 3　　(E) 1 : 2

Solution

Using the formula for the volume of a cylinder $V=\pi r^2h$, we get that the volume of Bill's can is $4^2\times15\times\pi$, and that the volume of David's can is $5^2\times8\times\pi$. Now, we divide the volume of Bill's can by the volume of David's can, so we get 6 : 5 (A).

解 析

利用圆柱体的体积公式 $V=\pi r^2h$，可得到比尔罐子的体积是 $4^2\times15\times\pi$，而大卫罐子的体积是 $5^2\times8\times\pi$。现在，我们用比尔罐子的体积除以大卫罐子的体积，因此比例为 6∶5，故选 (A)。

Example 4

Isabella uses one-foot cubical blocks to build a rectangular fort that is 12 feet long, 10 feet wide, and 5 feet high. The floor is one-foot-thick, and the four walls are all 2 feet thick. How many blocks does the fort contain?

例题 4

伊莎贝拉用 1 英尺长的立方体积木搭建了一个长 12 英尺、宽 10 英尺、高 5 英尺的长方形堡垒。地面是 1 英尺厚，而四面墙都是 2 英尺厚。这个堡垒能容纳的体积是多少？

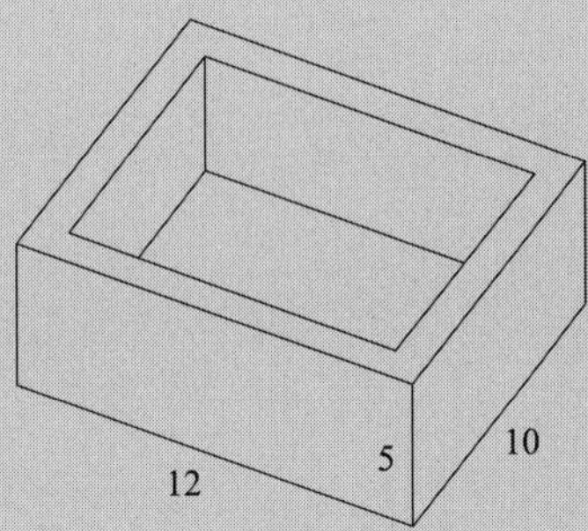

(A) 408　　(B) 200　　(C) 360　　(D) 320　　(E) 280

Solution

This question is testing on the formula for the volume of a cube. The volume of the outer cube is $12\times10\times5=600$. the volume of the inner cube is $(12-4)\times(10-4)\times(5-1)=192$. Therefore, the answer is $600-192=\text{(A)}408$.

解　析

这题考查的是正方体的体积公式。外侧的立方体体积是 $12\times10\times5=600$。内侧的立方体体积是 $(12-4)\times(10-4)\times(5-1)=192$。因此，答案是 $600-192=408$，故选 (A)。

Example 5

Any three vertices of the cube *PQRS-WVUT*, shown in the figure below, can be connected to form a triangle. (For example, vertices *P*, *Q*, and *R* can be connected to form isosceles $\triangle PQR$.) How many of these triangles are equilateral and contain *P* as a vertex?

例题 5

如图所示，正方体 $PQRS\text{-}WVUT$ 的任意三个顶点都可以连接成一个三角形。(例如，顶点 P，Q 和 R 可以连接成等腰三角形 PQR。) 在这些三角形中，有多少个是等边三角形并包含 P 作为顶点?

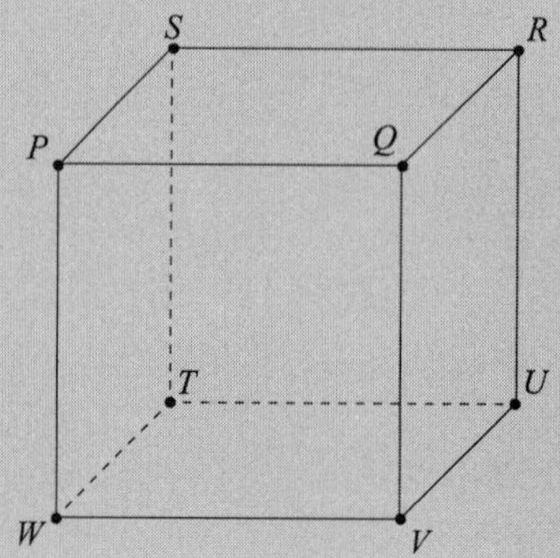

(A) 0 (B) 1 (C) 2 (D) 3 (E) 6

Solution

The only equilateral triangles that can be formed are through the diagonals of the faces of the square. From P you have 3 possible vertices that are possible to form a diagonal through one of the faces. Therefore, there are 3 possible triangles. So the answer is (D) 3.

解 析

只有通过正方形各面的对角线才能形成等边三角形。从点 P 出发，有 3 个顶点可以通过其中一个面形成对角线。因此，有 3 个可能的三角形。所以答案是 3，故选 (D)。

Example 6

In the cube $ABCD\text{-}EFGH$ with opposite vertices C and E, J and I are the midpoints of edges FB and HD respectively. Let R be the ratio of the area of the cross-section $EJCI$ to the area of one of the faces of the cube. What is R^2?

例题 6

在顶点 C 和 E 相对的立方体 $ABCD\text{-}EFGH$ 中，J 和 I 分别是边 FB 和 HD 的

中点。设 R 是截面 $EJCI$ 的面积与立方体一个面的面积之比。问 R^2 是多少？

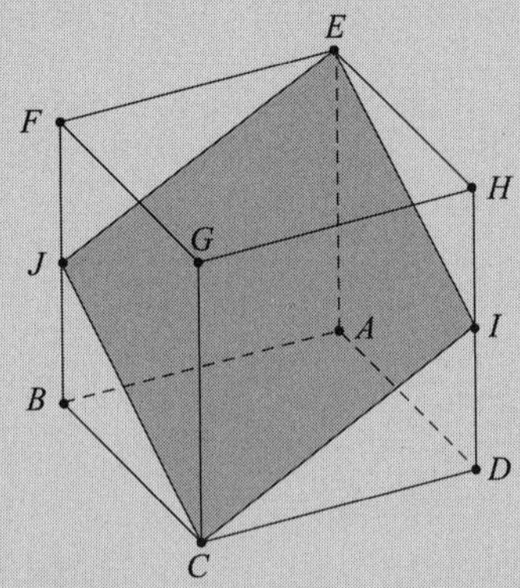

(A) $\frac{5}{4}$ (B) $\frac{4}{3}$ (C) $\frac{3}{2}$ (D) $\frac{25}{16}$ (E) $\frac{9}{4}$

Solution

Note that $EJCI$ is a rhombus by symmetry. Let the side length of the cube be 1. By the Pythagorean theorem, $EC=\sqrt{3}$, and $JI=\sqrt{2}$. Since the area of a rhombus is half the product of its diagonals, the area of the cross section is $\frac{\sqrt{6}}{2}$. This gives $R=\frac{\sqrt{6}}{2}$. Thus, $R^2=\frac{6}{4}=\text{(C)}\frac{3}{2}$.

解 析

注意，根据对称性，$EJCI$ 是一个菱形。假设立方体的边长为 1。根据勾股定理，$EC=\sqrt{3}$，$JI=\sqrt{2}$。由于菱形的面积是其对角线积的一半，所以横截面的面积为 $\frac{\sqrt{6}}{2}$。由此得出 $R=\frac{\sqrt{6}}{2}$。因此，$R^2=\frac{6}{4}=\frac{3}{2}$，故选 (C).

Example 7

The figure below shows a polygon $ABCDEFGH$, consisting of rectangles and right triangles. When cut out and folded on the dotted lines, the polygon forms a triangular prism. Suppose that $AH=9$, $EF=7$ and $GH=17$. What is the volume of the prism?

例题 7

下图是由矩形和直角三角形组成的多边形 $ABCDEFGH$。按虚线剪开并折叠后，多边形形成一个三棱柱。假设 $AH=9$，$EF=7$，$GH=17$。问棱柱的体积是多少?

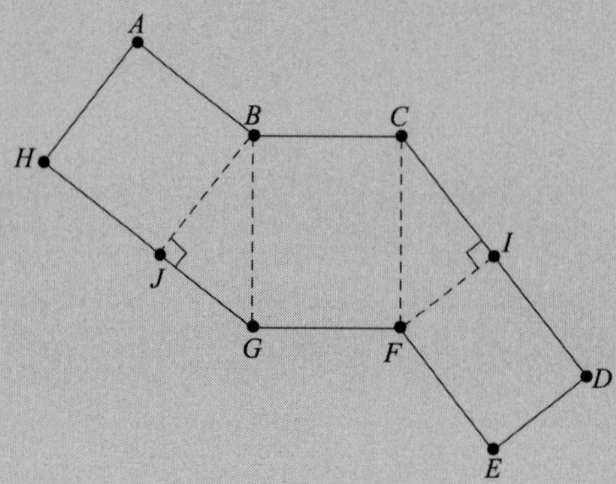

(A) 378　　(B) 415　　(C) 190　　(D) 310　　(E) 315

Solution

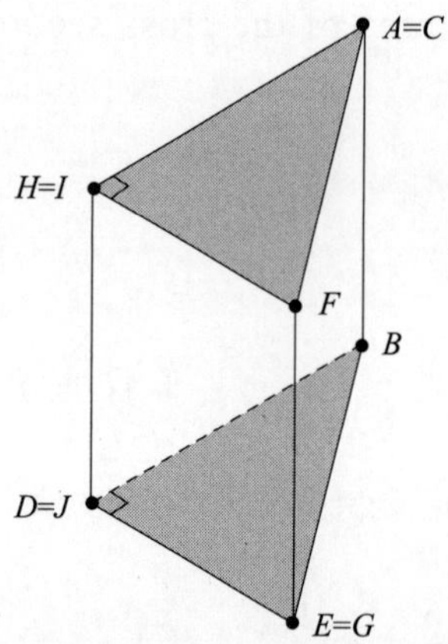

This question is about the formula for the volume of a triangular prism. After fold, $BJ = CI = 9$ and $FG = BC = HJ = 7$, so $DE = 17 - 7 = 10$. The area of $\triangle BJG$ is $\frac{1}{2}\times(9\times 10)=45$. If we let $\triangle BJG$ be the base, then the height is $HJ = 7$. So, the volume is $45\times 7 =$ (E) 315.

解　析

这题考查的是三棱柱的体积公式。折叠后的图像如图所示，$BJ = CI = 9$，$FG =$

$BC = HJ = 7$，所以 DE=17−7=10。$\triangle BJG$ 的面积是 $\frac{1}{2}\times(9\times10)=45$。则以 $\triangle BJG$ 为底，高为 $HJ=7$ 的三棱柱体积为 45×7=315，故选 (E)。

PRACTICE 练习

Problem 1

Jamie counted the number of edges of a cube, Jimmy counted the numbers of corners, and Judy counted the number of faces. They then added the three numbers. What was the resulting sum?

问题 1

杰米数正方体的边数，吉米数顶点数，朱迪数面数。然后他们把这 3 个数字相加。得出的总和是多少?

(A) 12 (B) 16 (C) 20 (D) 22 (E) 26

Problem 2

For his birthday, Cindy got a box that held 100 jellybeans when filled to capacity. A few weeks later, Emma got a larger bo0x full of jellybeans. Her box was 3 times as high, 3 times as wide and 2 times as long as Cindy's. Approximately, how many jellybeans did Emma get?

问题 2

辛迪的生日礼物是一个装有 100 颗软糖的盒子。几周后，爱玛得到了一个更大的装满软糖的盒子。她的盒子比辛迪的高 3 倍，宽 3 倍，长 2 倍。爱玛大约得到了多少颗软糖?

(A) 500 (B) 900 (C) 800 (D) 2700 (E) 1800

Problem 3

A rectangular box having no top is four times as long as it is wide and two times as tall as it is wide. If its volume is 512 cm^3, then its surface area in square centimeters is?

问题 3

一个没有顶部的长方体的长度是其宽度的 4 倍，高度是其宽度的 2 倍。如果它的体积是 512 立方厘米，那么它的表面积是多少平方厘米?

(A) 416　　(B) 384　　(C) 448

(D) 320　　(E) none of the above（以上都不是）

Problem 4

Suppose two pyramids are similar and the surface area of the larger is 27 times the volume of the smaller. How many times the surface area of the smaller pyramid is the surface area of the larger pyramid?

问题 4

假设两个金字塔相似且较大金字塔的体积是较小金字塔的 27 倍。那么较大金字塔的表面积是较小金字塔表面积的多少倍？

(A) 18　　(B) 3　　(C) 6　　(D) 27　　(E) 9

Problem 5

In the diagram, each edge of the cube has length 16. E, F, and G are the midpoints of CD, BD, and DI. Find the volume of the pyramid D-EFG.

问题 5

在图中，正方体的每条边的长度都是 16。E，F 和 G 分别是 CD、BD 和 DI 的中点。求三棱锥 D-EFG 的体积。

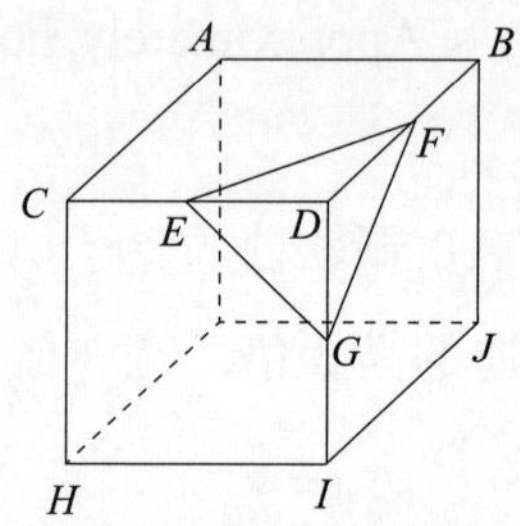

(A) 512　　(B) 256　　(C) $256\sqrt{3}$

(D) $\dfrac{256}{3}$　　(E) 86

Problem 6

A 15-inch by 15-inch piece of cardboard has a 3-inch by 3-inch square cut out of each corner. The sides are then folded up to form a box. What is the volume, in cubic inches, of the box?

问题 6

在一张 15 英寸 × 15 英寸的硬纸板上，每个角都剪出一个 3 英寸 × 3 英寸的正方形。然后将两边折叠起来形成一个盒子。这个盒子的体积是多少立方英寸?

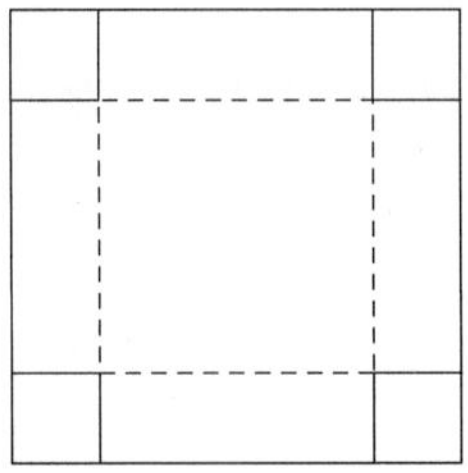

(A) 144 (B) 135 (C) 675 (D) 432 (E) 243

Problem 7

When the radius of the base of a cylinder was unchanged but its height was as seven times as the original height, the total surface area of the new cylinder was five times as large as the original. What was the ratio of the radius to the height of the original cylinder?

问题 7

当一个圆柱体的底面半径不变而高度变为原来的 7 倍时，新圆柱体的总表面积是原来圆柱体的 5 倍。原来圆柱的半径和高的比是多少?

(A) 1 : 1 (B) 2 : 1 (C) 1 : 2 (D) 3 : 1 (E) 1 : 4

Problem 8

Suppose the side, front, and bottom faces of a right rectangular solid have areas of 12 cm^2, 20 cm^2, and 15 cm^2 respectively, then the volume is:

问题 8

假设一个直角长方体的侧面、正面和底面的面积分别为 12 平方厘米，20 平方厘米和 15 平方厘米，那么它的体积是多少立方厘米?

(A) 24 (B) 60 (C) 12

(D) 47 (E) none of these 以上都不是

Problem 9

A triangular pyramid with vertices T, U, V and Q is removed from the solid cube shown. How many edges does the remaining solid have?

问题 9

从如图所示的实心正方体中移除一个顶点为 T，U，V 和 Q 的三棱锥。剩下的立体图形有多少条棱？

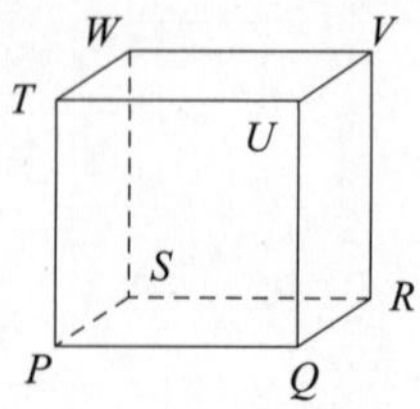

(A) 4　　(B) 6　　(C) 8　　(D) 10　　(E) 12

Problem 10

What is the remaining volume in cubic centimeters if a cylinder with a radius of 5 cm is removed from a rectangular solid whose sides measure 10 cm × 10 cm × 20 cm? Express your answer in terms of π.

问题 10

从一个 10 厘米 × 10 厘米 × 20 厘米的长方体中取出一个半径为 5 厘米的圆柱体，剩余的体积是多少立方厘米？用 π 表示答案。

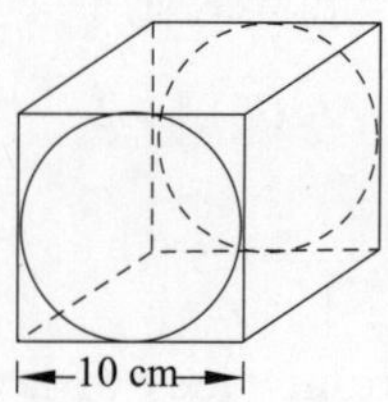

(A) $2000-500\pi$　　(B) $2000-250\pi$　　(C) $1000-250\pi$

(D) $2000+500\pi$　　(E) 25π

Problem 11

Fourteen white cubes are put together to form the figure shown. The complete surface of the figure, excluding the bottom, is painted black. The figure is then separated into individual cubes. How many of the individual cubes have exactly three black faces?

问题 11

将 14 个白色立方体放在一起组成一个立体图形。整个立体图形的表面，除底部外，都被涂成黑色。然后该图形被分成单独的立方体。有多少个立方体恰好有三个黑色面？

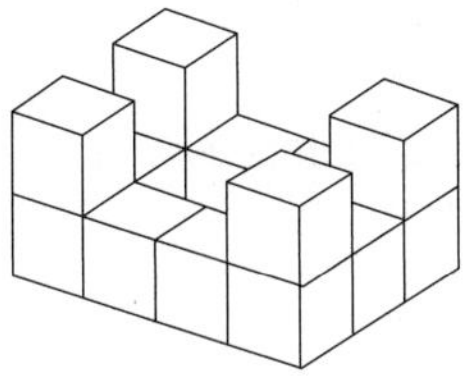

(A) 5 (B) 6 (C) 7 (D) 8 (E) 0

Problem 12

A wire is wrapped around a cylinder forming a helix as in the picture. If the wire only goes around the cylinder once with the height 30π cm, and the diameter 40 cm, find the length of the wire.

问题 12

如图所示，将一根钢丝缠绕在圆柱体上形成螺旋。如果钢丝仅绕圆柱体一周，且圆柱体的高度为 30π，直径为 40 厘米，求钢丝的长度。

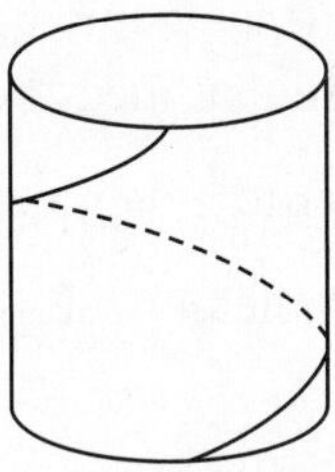

(A) 50π (B) 50 (C) $\sqrt{50\pi}$ (D) $\pi\sqrt{50}$ (E) 2500

Problem 13

A cube with sides of length 10 cm has three face diagonals drawn as shown to form the edges of a new (6-edged) solid. What is the number of cubic centimeters in the volume of the new solid?

问题 13

边长为 10 厘米的立方体按如图所示绘制的三个面对角线形成了新的（6 边）立体图形。新立体图形的体积是多少立方厘米？

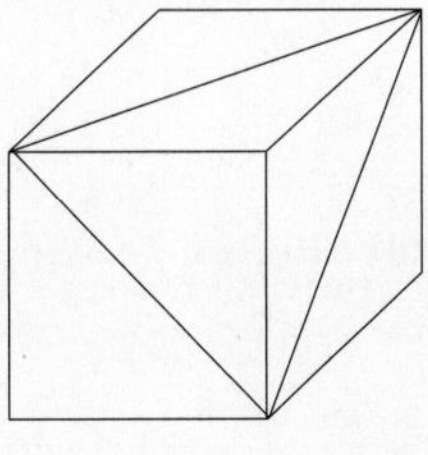

(A) $\frac{500}{3}$　　(B) 500　　(C) $\frac{1000}{3}$　　(D) $10\sqrt{2}$　　(E) $100\sqrt{2}$

Problem 14

A cube with 3-inch edges is to be constructed from 27 smaller cubes with 1-inch edges. 17 of the cubes are colored white and 10 are colored blue. If the 3-inch cube is constructed to have the biggest possible blue surface area showing, what fraction of the surface area is blue?

问题 14

一个边长为 3 英寸的立方体由 27 个较小的有 1 英寸边缘的立方体组成。其中 17 个立方体是白色的，10 个是蓝色的。如果使 3 英寸的立方体显示最大的蓝色表面积，那么蓝色表面积占总表面积的几分之几？

(A) $\frac{14}{27}$　　(B) $\frac{14}{27}$　　(C) $\frac{13}{27}$　　(D) $\frac{29}{54}$　　(E) $\frac{13}{54}$

Problem 15

An aquarium has a rectangular base that measures 40 cm by 50 cm and has a height of 60 cm. The aquarium is filled with water to a depth of 48 cm. A rock with volume 2000 cm^3 is then placed in the aquarium and completely submerged. By how many centimeters does the water level rise?

问题 15

水族箱有一个长方形底座，尺寸为 40 厘米 × 50 厘米，高 60 厘米。水族箱注水深度为 48 厘米。然后将一块体积为 2000 立方厘米的石头放入水族箱中并完全浸没。问水位上升了多少厘米？

(A) 1　　(B) 0.5　　(C) 1.5　　(D) 0.25　　(E) 0.2

Chapter 12 Statistics

第12章 统计学

12.1 Basic Knowledge 基础内容

12.1.1 Mean (Arithmetic mean) 平均数（算术平均数）

The mean of a set of data consisting of n numbers is the sum of numbers in the set divided by n.

The mean of the numbers $x_1, x_2, \cdots, x_n$ denoted by $\bar{x}$ and read "x bar", is given by $\bar{x} = \frac{x_1 + x_2 + \cdots + x_n}{n}$.

The mean does not have to be a member of the set of data.

一组由 n 个数构成的集合的平均数是该组数据的和除以 n。

该组数 $x_1, x_2, \cdots, x_n$ 的平均数，记为 $\bar{x}$（读作"x bar"），其计算公式为 $\bar{x} = \frac{x_1 + x_2 + \cdots + x_n}{n}$。

平均数不一定是数组中出现的数。

Example 1

The bar chart shows the number of days in a week that students visited the gym. What was the mean number of days of exercise in this week, rounded to the nearest hundredth?

例题 1

柱状图显示了学生一周内去健身房锻炼的天数，这些学生这一周锻炼的平均天数为多少(四舍五入到小数点后两位)?

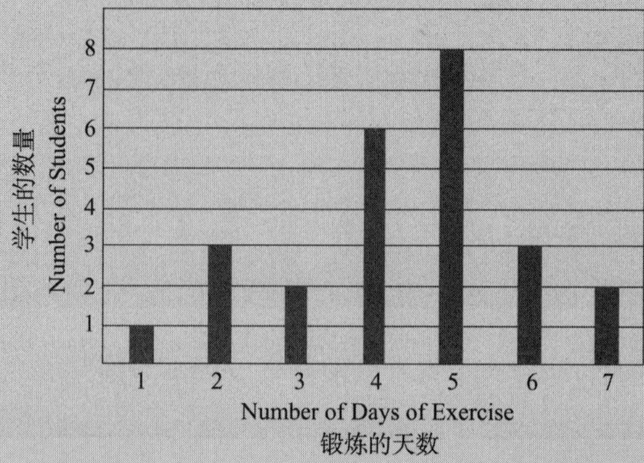

(A) 3.50 (B) 3.57 (C) 4.36 (D) 4.50 (E) 5.00

Solution

According to the definition of mean value, we need to find the total number of days of exercise, and the number of students.

According to the diagram, the total number of days of exercise is

$$1\times1+2\times3+3\times2+4\times6+5\times8+6\times3+7\times2=109$$

The total number of students is

$$1+3+2+6+8+3+2=25$$

Therefore, the mean is

$$\frac{109}{25}=4.36$$

The answer is (C).

解 析

根据平均值的定义,需要计算运动的总天数和学生的总人数。

根据柱形图,运动的总天数是

$$1\times1+2\times3+3\times2+4\times6+5\times8+6\times3+7\times2=109$$

学生的总人数是

$$1+3+2+6+8+3+2=25$$

因此,平均值是

$$\frac{109}{25}=4.36$$

所以答案是 (C)。

Example 2

Four numbers are written in a row. The average of the first two is 11, the average of the middle two is 16, and the average of the last two is 20. What is the average of the first and last of the numbers?

例题 2

4 个数字排成一列。前两个数的平均数是 11，中间两个数的平均数是 16，最后两个数的平均数是 20。第一个和最后一个数的平均数是多少？

(A) 14 (B) 15 (C) 16 (D) 17 (E) 18

Solution

Let a, b, c, and d be the four numbers. Since the average of a and b is 11 and that of b and c is 16, we know $a+b=22$ and $b+c=32$ By subtracting the two equations, we have $a-c=-10$. Since the average of c and d is 20, we also get $c+d=40$. By adding the last two equations together, we have $a+d=30$. So, the average of a and d is (B) 15.

解 析

设这 4 个数为 a、b、c 和 d。由题意可知，a 和 b 的平均值是 11，b 和 c 的平均值是 16，即 $a+b=22$，$b+c=32$。将这两个等式相减，得到 $a-c=-10$。又因为 c 和 d 的平均值是 20，即 $c+d=40$。将最后两个等式相加，得到 $a+d=30$。因此，a 和 d 的平均值是 15，故选 (B)。

12.1.2 Median 中位数

The median is the number in the middle when the data are arranged in order from least to greatest. When there are two middle numbers, the median is their mean.

Median is also the value that divides a group of numbers into two parts, with half of the numbers smaller than the median and half greater than it.

Steps to find the median of a set of numbers:

- Arrange all the n in the set in numerical order from smallest to greatest.

- The middle number is the median if n is odd.
- The mean of two middle numbers is the median if n is even.

中位数是在将数字按从小到大的顺序排列时位于中间的数，如果中间有两个数，则中位数是这 2 个数字的平均数。

中位数将一组数分成两部分，其中一部分数小于中位数，另一部分数大于中位数。

找到一组数的中位数的步骤如下：

- 将数组中所有 n 个数按从小到大的顺序排列。
- 如果 n 是奇数，则中位数为最中间的那个数。
- 如果 n 是偶数，则中位数是中间两个数的平均数。

Example 3

In a recent music competition, the judges scored 20 participants' performance. The dot plot below displays the distribution of scores given out of 100 points. After reviewing the scores, the head judge realized that an error had occurred in calculating the scores for a few participants, resulting in an incorrect score for some. To correct this, they decided to add 5 extra points to those affected, which shifted the median score to 85. What is the minimum number of participants who received the extra points?

(Note that the median score equals the average of the 2 scores in the middle if the 20 scores are arranged in increasing order.)

例题 3

在最近的一场音乐比赛中，评委们对 20 位参赛者的表演进行了打分。下面的点阵图显示了满分 100 分的分数分布情况。在复查得分后，主评委意识到在计算部分参赛者的分数时出现了错误，导致一些参赛者的分数不正确。为了纠正这一错误，评委们决定为受影响的参赛者额外加 5 分，这使得中位数得分变为了 85 分。请问至少有多少名参赛者得到了额外加分？

（注意，如果分数按从小到大的顺序排列，那么中位数等于 20 个数的中间 2 个数的平均值。）

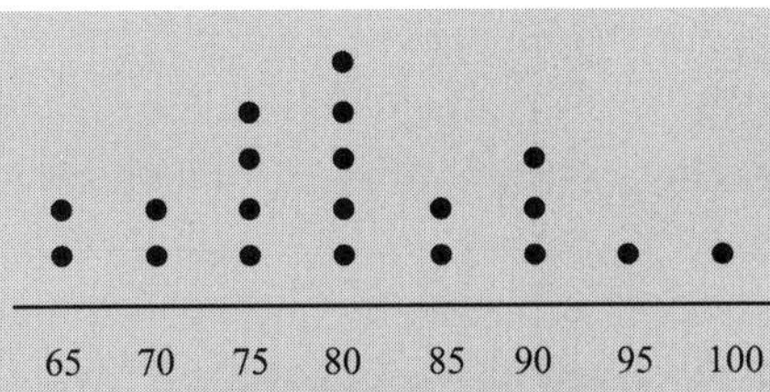

(A) 2　　(B) 3　　(C) 4　　(D) 5　　(E) 6

Solution

According to the dot plot, the 10th and 11th scores were 80 before the correction, so the median was also 80. To increase the median to 85, we should make the 10th and 11th score 85. This means we need to add 5 extra points to at least 4 participants, the 10th, 11th, 12th, and 13th. The answer is (C).

解　析

由点阵图可知，第 10 个和第 11 个分数在修改之前都是 80，即中位数是 80。为了将中位数提高到 85，必须将第 10 名和第 11 名的分数改为 85 分。所以至少第 10 位、第 11 位、第 12 位和第 13 位参赛者，共 4 人需要获得额外 5 分加分。故答案是 (C)。

Example4

Mark is in the 5th grade and weighs 96 pounds. His quadruplet sisters are tiny babies and weigh 4, 4, 7, and 9 pounds. Which is greater, the average (mean) weight of these five children or the median weight, and by how many pounds?

例题 4

马克上五年级，体重 96 磅。他的四胞胎妹妹们分别重 4 磅、4 磅、7 磅和 9 磅。5 个孩子体重的平均数和中位数哪个更大，相差多少?

(A) median, by 20 中位数，差 20

(B) median, by 17 中位数，差 17

(C) average, by 15 平均数，差 15

(D) average, by 17 平均数，差 17

(E) average, by 20 平均数，差 20

Solution

By the definition of mean, the average weight of the children is

$$(96+4+4+7+9)\div 5=24$$

Since there are 5 weights, the median weight is the 3rd weight in ascending order, 7. So, the average weight is 17 pounds more than the median weight. The answer is (B).

解 析

根据平均值的定义，孩子们体重的平均值是

$$(96+4+4+7+9)\div 5=24$$

由于有 5 个孩子，所以体重的中位数按递增排列的第三个数字，即 7。所以平均值比中位数更大，两者相差 17 磅。答案是 (B)。

12.1.3 Mode 众数

The mode of a set of data is the most frequently occurring number or numbers in the data set.

The mode is always a member of the set of data. The mode is not necessarily unique. Some sets of numbers have more than one most frequently occurring value. A set of numbers with two modes is said to be bimodal while a set with more than two modes may be described as multimodal.

一个数组的众数是该集合中出现次数最多的一个或多个数。

众数一定是集合中的某个数。但众数不一定是唯一的。有些集合有多个值会频繁出现。若一组数有两个众数，则称为双峰分布，若有多个众数，则称为多峰分布。

Example 5

Find the positive difference of the mode and the median of this set of scores: 28, 22, 26, 35, 20, 35, 15, 18.

例题 5

找出该组数据众数和中位数的差值：28, 22, 26, 35, 20, 35, 15, 18。

(A) 10　　(B) 11　　(C) 12　　(D) 14　　(E) 13

Solution

According to the definition, mode is the number mostly frequently occurring, so the mode is 35. Since there are 8 numbers here, the median is the average of the 4th and 5th in the increasing order, which is $\frac{22+26}{2}=24$. Thus, the positive difference is $|35-24|=11$. The answer is (B).

解 析

根据定义，一组数的众数为出现次数最多的那个，因此众数为 35。数组中有 8 个数，因此中位数应为从小到大排列后第 4 个数和第 5 个数的平均数，即 $\frac{22+26}{2}=24$。因此，二者的差 $|35-24|=11$，答案是 (B)。

Example 6

A company is evaluating the efficiency of their assembly lines. The production rate, measured in units per hour for each day of the week, is given as 24，28，30，30，32，34，34. Which statement about the mean, median and mode is true?

例题 6

一家公司正在评估其装配线的效率，这条装配线一周的效率，以每小时生产的产品件数衡量，分别为 24、27、28、31、32、34、34。关于其效率的平均数、中位数和众数，以下哪个说法是正确的?

(A) median (中位数)<mean (平均数)<mode (众数)

(B) mean (平均数)<mode (众数)<median (中位数)

(C) mean (平均数)<median (中位数)<mode (众数)

(D) median (中位数)<mode (众数)<mean (平均数)

(E) mode (众数)<median (中位数)<mean (平均数)

Solution

According to the definitions, the mean is

$$(24+27+28+31+32+34+34)\div 7=30$$

The median is the 4th number, 31, and the mode the number with the most occurrences, which is 34. So, mean < median < mode. The answer is (C).

解 析

根据定义，平均数是 $(24+27+28+31+32+34+34)\div 7=30$，中位数是第4个数字31，众数是出现最多的那个数字，即34。所以，平均数<中位数<众数，故选(C)。

12.1.4 Range 极差

The range of a set of numbers is the difference between the least and the greatest number in the set.

一组数的极差是该集合中最大数与最小数的差值。

Example 7

Billy's basketball team scored the following points over the course of the first 11 games of the season: 43, 48, 54, 54, 59, 59, 59, 62, 65, 66, 74. If his team scores 41 in the 12th game, which of the following statistics will show an increase?

例题 7

比利的篮球队在赛季前11场比赛中的得分如下：43、48、54、54、59、59、59、62、65、66、74。如果他的球队在第12场比赛中得到41分，以下哪一项统计数据将会增加？

(A) range 极差

(B) median 中位数

(C) mean 平均数

(D) mode 众数

(E) mid-range 中程数

Solution

Since 41 is the smallest of the twelve scores, adding it will increase the range by the definition of range. The new median is the average of 59 and 59, since they are the 6th and 7th scores in the ascending order, which is the same as the previous one. Previously the mean is

$\frac{643}{11}$, which is greater than 41. According to the formula of mean, adding 41 will decrease it. Since the occurrence of 59 is 3 and 41 is a new number, the mode will stay the same as it is the number of greatest occurrences. Finally, since the mid-range is the average of the greatest and the smallest numbers, it will decrease as now the smallest is not 43 but 41. Therefore, the answer is (A).

解析

由于41是12个数中最小的，根据极差的定义，添加它会导致极差的上升。在新的数组中从小到大第6个和第7个数字都是59，因此新的中位数59与之前的相同。之前的平均数为$\frac{643}{11}$，大于41。根据平均数计算公式，添加41将降低平均数。由于59出现了3次，而41是一个新数字，在新的数组中59依旧是出现最多次的，所以众数将保持不变。最后，由于中程数是最大值和最小值的均值，新的中程数将比原来小，因为现在最小值不是43而是41，所以答案应该选(A)。

Example 8

Two integers are inserted into the list 13, 13, 18, 21, 38 to double its range. The mode and median remain unchanged. What is the maximum possible sum of two additional numbers?

例题 8

在数列13、13、18、21、38中插入两个整数，使其极差加倍。且众数和中位数保持不变。这两个数的和最大是多少？

(A) 76　(B) 77　(C) 78　(D) 80　(E) 81

Solution

According to the definition of range, the current range is $38-13=25$. So, the new range is 50. Also by the definition, to make the median unchanged, one of the integers must be less than or equal to 18 and the other greater or equal to 18. So, we can assume the numbers are a and b such that $a \leqslant 18 \leqslant b$. To make the mode unchanged, we know $a<18<b$ and $b \neq 21,38$.

If $a \leqslant 13$, then we have $b-a=50$ and $a+b=50+2a$. So, we can maximize the sum by making $a=13$ to have the maximum as 76 in this case.

If $13<a<18$, then we know $b-13=50$ and $b=63$. Thus, the sum is maximized when a is maximized, which is 17. Therefore, the maximum is 80 in this case.

Since 76 < 80, the answer is (D).

解析

根据极差的定义，当前的极差是$38-13=25$，所以新的极差是50。根据中位数的定义，为了使中位数保持不变，其中的一个整数应当小于等于18，而另一个应当大于等于18。设这两个数字分别是a和b，并且它们满足$a\leqslant 18\leqslant b$。为了使众数保持不变，$a<18<b$以及$b\neq 21$且$b\neq 38$。

若$a\leqslant 13$，则$b-a=50$且$a+b=50+2a$。因此，当$a=13$时，它们的和最大，这种情况下的最大值是76。

若$13<a<18$，则$b-13=50$，即$b=63$。当a最大为17时，总和最大。在这种情况下，最大值是80。

又因为76 < 80，所以答案应该选(D)。

Example 9

The mean, median, and unique mode of the positive integers 13, 14, 15, 16, 16, 17, and x are all equal. What is the value of x?

例题 9

数组13，14，15，16，16，17，x的平均数、中位数和众数都相等，且众数只有一个。x的值是多少？

(A) 15 (B) 16 (C) 17 (D) 21 (E) 22

Solution

According to the definition of mode, the mode is 6 if x is not in the list. Since the mode is unique, either $x = 16$ or $x\neq 13, 14, 15, 16, 17$. So, we know the mode, median, and mean are all 16. Therefore, we have $13+14+15+16+16+17=91+x = 112$ and $x = 21$. The answer should be (D).

解 析

根据众数的定义，在没有 x 时，数组的众数是 16。由于数组的众数只有一个，所以要么 $x=16$，要么 x 不取 13，14，15，16，17。因此，众数、中位数和平均数都是 16，即 $13+14+15+16+16+17=91+x=112$，可得 $x=21$。答案应该选 (D)。

PRACTICE 练习

Problem 1

The mean of the list $n+10$, $n+13$, $n+14$, $n+15$, $n+16$, $n+18$, $n+20$, $n+22$, $n+25$ is 21. What is the median?

问题 1

数列 $n+10$，$n+13$，$n+14$，$n+15$，$n+16$，$n+18$，$n+20$，$n+22$，$n+25$ 的平均数是 21。问中位数是多少？

(A) 14 (B) 16 (C) 17 (D) 20 (E) 21

Problem 2

The average age of 30 students is 12. The average age of 50 of their parents is 36. What is the average age of all of these parents and students?

问题 2

30 名学生的平均年龄是 12 岁，50 名学生家长的平均年龄是 36 岁。所有这些家长和学生的平均年龄是多少？

(A) 23 (B) 24.25 (C) 27 (D) 28 (E) 29.25

Problem 3

The average age of 6 people in a room is 25 years old. A 20-year-old person leaves the room. What is the average age of the remaining people?

问题 3

房间里 6 个人的平均年龄是 25 岁。一个 20 岁的人离开了房间。剩下人的平均年龄是多少？

(A) 23 (B) 24 (C) 25 (D) 26 (E) 27

Problem 4

The average of the five numbers in a list is 27. The average of the first two numbers is 24. What is the average of the last three numbers?

问题 4

一个列表中 5 个数的平均值是 27，前 2 个数的平均值是 24。问后 3 个数的平均值是多少？

(A) 26 (B) 27 (C) 28 (D) 29 (E) 30

Problem 5

The mean, median, and mode of the 7 data values 70, 110, x, 50, 60, 210, 100 are all equal to x. What is the value of x?

问题 5

7 个数 70，110，x，50，60，210，100 的平均数、中位数和众数都等于 x。问 x 的值是多少？

(A) 60 (B) 70 (C) 85 (D) 100 (E) 110

Problem 6

Of the 48 students in Richelle's class, 18 prefer chocolate pie, 12 prefer apple, and 6 prefer blueberry. Half of the remaining students prefer cherry pie and half prefer lemon. For Richelle's pie graph showing this data, how many degrees should she use for cherry pie?

问题 6

雷切尔班上有 48 名学生，18 名喜欢巧克力派，12 名喜欢苹果派，6 名喜欢蓝莓派。剩下的一半学生喜欢樱桃派，另一半喜欢柠檬派。雷切尔应该在饼状图中使用多少度的角代表喜欢樱桃派的学生？

(A) 10 (B) 25 (C) 30 (D) 45 (E) 60

Problem 7

Mary's parents have agreed to buy her tickets to see her favorite band if she spends an average of 8 hours per week helping around the house for 7 weeks. For the first 6 weeks she helps around the house for 6, 9, 5, 10, 8, and 7 hours. How many hours must she work for the final week to earn the tickets?

问题 7

玛丽的父母同意为她买票去看她最喜欢的乐队演出，条件是她在 7 周内每周平均花 8 小时在家帮忙。前 6 周她分别帮忙了 6 小时、9 小时、5 小时、10 小时、8 小时和 7 小时。她最后一周必须帮忙多少小时才能获得门票？

(A) 9 (B) 10 (C) 11 (D) 12 (E) 13

Problem 8

An iterative average of the numbers 11, 12, 13, 14, and 15 is computed the following way. Arrange the five numbers in some order. Find the mean of the first two numbers, then find the mean of that with the third number, then the mean of that with the fourth number, and finally the mean of that with the fifth number. What is the difference between the largest and smallest possible values that can be obtained using this procedure?

问题 8

计算数字 11、12、13、14 和 15 的迭代平均数的方法如下。将 5 个数字按一定顺序排列。首先找到前两个数字的平均数，然后找到它与第 3 个数字的平均数，再找到这个平均数与第 4 个数字的平均数，最后找到它与第 5 个数字的平均数。问这个方法所能获得的最大值和最小值之间的差是多少？

(A) $\frac{31}{16}$ (B) 2 (C) $\frac{17}{8}$ (D) 3 (E) $\frac{17}{8}$

Problem 9

How many subsets of two elements can be removed from the set $\{1, 2, 3, 4, 5, 6, 7, 8, 9\}$ so that the mean (average) of the remaining numbers is 5?

问题 9

从集合 $\{1, 2, 3, 4, 5, 6, 7, 8, 9\}$ 中可以移除多少个包含两个元素的子集，使得剩下数字的平均数是 5？

(A) 1 (B) 2 (C) 3 (D) 4 (E) 5

Problem 10

The average of the numbers $1, 2, 3, \cdots, 48, 49$, and x is $2x$. What is x?

问题 10

数列 1，2，3，…，48，49，x 的平均数是 $2x$，问 x 是多少？

(A) $\frac{1220}{99}$ (B) $\frac{1225}{99}$ (C) $\frac{1230}{99}$ (D) $\frac{1235}{99}$ (E) $\frac{1240}{99}$

Problem 11

Let R be a set of nine distinct integers. Six of the elements are 12, 13, 14, 16, 19, and 24. What is the number of possible values of the median of R?

问题 11

设 R 为一个包含 9 个不同整数的集合。其中 6 个数为 12、13、14、16、19 和 24。问 R 的中位数可能有多少个不同的值？

(A) 4 (B) 5 (C) 6 (D) 7 (E) 8

Problem 12

Find the positive difference between the mean and the mode of this set of data: 11, 11, 12, 12, 12, 13, 13, 14, 19.

问题 12

找出数组 11、11、12、12、12、13、13、14、19 的平均数与众数之差。

(A) 3 (B) 2 (C) 1 (D) 0 (E) 4

Problem 13

The average (mean) of 60 numbers is 40, and the average of 40 other numbers is 60. What is the average of all numbers?

问题 13

60 个数的平均数是 40，另外 40 个数的平均数是 60。问所有数的平均数是多少？

(A) 40 (B) 42 (C) 44 (D) 46 (E) 48

Problem 14

A sample consisting of five integral observations has an arithmetic mean of 22 and a median of 24. The smallest value that the range (largest observation minus smallest) can assume for such a sample is?

问题 14

5 个整数样本的平均值是 22，中位数是 24。问样本极差（最大观测值减去最小观测值）的最小值是多少？

(A) 2 (B) 3 (C) 5 (D) 7 (E) 10

Problem 15

The mean of three numbers is 10 more than the least of the numbers and 15 less than the greatest. The median of the three numbers is 15. What is their sum?

问题 15

3 个数字的平均数比最小的数字大 10，比最大的数字小 15。这 3 个数字的中位数是 15，问它们的和是多少？

(A) 50 (B) 55 (C) 60 (D) 65 (E) 70

Problem 16

The mean of a set of five different positive integers is 20. The median is 23. What is the maximum possible value of the largest number in the set?

问题 16

一个由 5 个不同正整数构成的集合的平均数是 20，中位数是 23。问这 5 个整数中最大数的最大可能值是多少？

(A) 35 (B) 40 (C) 45 (D) 50 (E) 55

Problem 17

One day the Beverage Barn sold 355 cans of soda to 100 customers, and every customer bought at least one can of soda. What is the maximum possible median number of cans of soda bought per customer on that day?

问题 17

某一天饮料店向 100 名顾客卖出了 355 罐汽水，每位顾客至少买了一罐汽水。问当天每位顾客购买汽水数量的中位数最大可能值是多少？

(A) 4.5 (B) 5.0 (C) 5.5 (D) 6.0 (E) 6.5

Problem 18

A list of five positive integers has mean 13 and range 23. The mode and median are both 8. How many different values are possible for the second largest element of the list?

问题 18

5 个正整数的平均数是 13，极差是 23，众数和中位数都是 8。问其中第二大的数可以有多少种不同的值？

(A) 4 (B) 7 (C) 8 (D) 10 (E) 12

Problem 19

When 43 is appended to a list of integers, the mean is increased by 3; When 9 is appended to the enlarged list, the mean of the enlarged list is decreased by 2. How many integers were in the original list?

问题 19

当 43 添加到一组整数中时，平均数增加了 3；当 9 再添加到这组扩展后的整数中时，扩展后的数组的平均值减少了 2。问原数组中有多少个整数？

(A) 6 (B) 7 (C) 8 (D) 9 (E) 10

Problem 20

A list of integers has mode 42 and mean 32. The smallest number in the list is 20. The median m of the list is a member of the list. If the list member m were replaced by $m+10$, the mean and median of the new list would be 34 and $m+10$, respectively. If m were instead replaced by $m-8$, the median of the new list would be $m-4$. What is m?

问题 20

一组整数的众数是 42，平均数是 32，其中最小的数是 20。中位数 m 是这组数中的某一个。如果将 m 替换为 $m+10$，新数组的平均数和中位数分别为 34 和 $m+10$。如果将 m 替换为 $m-8$，新数组的中位数变为 $m-4$。问 m 的值是多少？

(A) 26 (B) 27 (C) 28 (D) 29 (E) 30

Chapter 13 Permutations and Combinations

第 13 章 排列与组合

13.1 Basic Knowledge 基础内容

13.1.1 Factorial 阶乘

The symbol! (factorial) is defined as follows:

$$0! = 1$$

and for integers $n \geqslant 1$,

$$n! = n \cdot (n-1) \cdot \cdots \cdot 1$$

符号（阶乘）“!”的定义如下：

$$0! = 1$$

对于整数 $n \geqslant 1$，有

$$n! = n \cdot (n-1) \cdot \cdots \cdot 1$$

$$2! = 2 \times 1 = 2$$

$$3! = 3 \times 2 \times 1 = 6$$

$$4! = 4 \times 3 \times 2 \times 1 = 24$$

$$5! = 5 \times 4 \times 3 \times 2 \times 1 = 120$$

$$6! = 6 \times 5 \times 4 \times 3 \times 2 \times 1 = 720$$

Example 1

For a positive integer n, the factorial notation $n!$ represents the product of the integers from n to 1. (For example, $6! = 6 \times 5 \times 4 \times 3 \times 2 \times 1$.) What value of N satisfies the following equation?

$$5! \times 14! = 8 \times N!$$

例题 1

对于正整数 n，阶乘符号 $n!$ 表示从 n 到 1 的整数乘积。（比如 $6! = 6\times5\times4\times3\times2\times1$ 。）求满足以下方程的 N 的值。

$$5! \times 14! = 8 \times N!$$

(A) 11　　(B) 12　　(C) 13　　(D) 14　　(E) 15

Solution

Expand the equation and have $5\times4\times3\times2\times1\times14! = 8\times N!$. We can divide both sides by 8 and get $N! = 5\times3\times14! = 15\times14! = 15!$. Thus the answer should be N = (E)15.

解　析

方程原式展开 $5\times4\times3\times2\times1\times14! = 8\times N!$，两边同时除以 8，得到 $N! = 5 \times 3 \times 14! = 15\times14! = 15!$ 。所以 N=15，答案是 (E)。

Example 2

If n is an even positive integer, the double factorial notation $n!!$ represents the product of all the even integers from 2 to n. For example, $8!! = 2\times4\times6\times8$. What is the units digit of the following sum?

$$2!! + 4!! + 6!! + \cdots + 1998!! + 2000!! + 2002!!$$

例题 2

假设 n 是偶数，双阶乘符号 $n!!$ 表示从 2 到 n 的所有偶数的乘积。例如 $8!! = 2\times4\times6\times8$ 。问下列和的个位数字是多少？

$$2!! + 4!! + 6!! + \cdots + 1998!! + 2000!! + 2002!!$$

(A) 0　　(B) 2　　(C) 4　　(D) 6　　(E) 8

Solution

Observe that if $n \geqslant 10$, then the unit digit of $n!!$ is always 0 by the definition of !!. We just need to find the unit digit of $2!!+4!!+6!!+8!!$, which is 2. So the answer is (B).

解 析

当 $n \geqslant 10$ 时，由双阶乘的定义可知，$n!!$ 的个位数总是 0。我们只要计算 $2!!+4!!+6!!+8!!$ 的个位数字即可，结果是 2。所以答案是（B）。

13.1.2 Permutation 排列

A permutation is an arrangement or a listing of things in which order is important.

(1) Take r elements each time from n distinct elements ($1 \leqslant r \leqslant n$).

Number of permutations: $P(n,r)=\dfrac{n!}{(n-r)!}$.

(2) n distinct objects can be permutated in $n!$ permutations.

We let $n=r$ in (1) to get $P(n,n)=n!$

排列是一种注重顺序的物体安排方式。

(1) 从 n 个不同元素中不放回地取出 r（$1 \leqslant r \leqslant n$）个元素，则排列数为 $P(n,r)=\dfrac{n!}{(n-r)!}$。

(2) n 个不同的元素可以有 $n!$ 种不同的排列。

我们可以令 (1) 中的 $n=r$，得到 $P(n,n)=n!$。

Example 3

In how many ways can the letters of the word CANDY be arranged?

例题 3

单词“CANDY”中的字母可以有多少种排列方式？

(A) 6 (B) 24 (C) 60 (D) 120 (E) 1

Solution

There are 5 different letters in CANDY. By the definition of $P(n,n)$, there are 5!=120 different ways to arrange them. So the answer is (D).

解　析

"CANDY"中有 5 个不同的字母。根据排列数 $P(n,n)$ 的定义，一共有有 $5!=120$ 种不同的排列方式，所以答案是 (D)。

Example 4

In how many ways can the letters in BFFKFFPFR be rearranged so that two or more Fs do not appear together?

例题 4

有多少种方式可以重新排列 BFFKFFPFR 中的字母使得两个或更多个 F 不会相邻？

(A) 1　　(B) 10　　(C) 12　　(D) 24　　(E) 120

Solution

There are one B, one K, one P, one R, and five Fs. To make neither Fs adjacent to each other, the only way is to insert other letters to positions between two Fs. There are 4 different such positions. So, the number of the arrangement is the number of the permutations of B, K, P, and R, which is $4!=24$. The answer is (D).

解　析

这串字母中有一个 B、一个 K、一个 P、一个 R 和 5 个 F。为了确保没有两个 F 是相邻的，唯一的方法是将其他的字母放到两个 F 之间的位置。由于有 4 个这样的位置，所以排放的数量与 B、K、P 和 R 的排列数相同，即 $4!=24$ 。因此答案是 (D)。

13.1.3　Combination 组合

A combination is an arrangement or a listing of things in which order is not important.

Definition: Let n, r be non-negative integers such that $0\leqslant r\leqslant n$. The symbol $\binom{n}{r}$ (read "n choose r") is defined and denoted by

$$\binom{n}{r}=\frac{P(n,r)}{P(r,r)}=\frac{n!}{(n-r)!r!}$$

Remember: $\binom{n}{0}=1$, $\binom{n}{1}=n$, and $\binom{n}{n}=1$.

Since $n-(n-r)=r$ we have $\binom{n}{r}=\binom{n}{n-r}$.

Unlike permutations, combinations are used when order does not matter. If we have n different elements, and it doesn't matter which order we take the elements, the number of ways to take m elements where $1\leqslant m\leqslant n$ is $\binom{n}{m}$.

组合是一种不注重顺序的物体安排方式。

定义：设 n 和 r 为非负整数且 $0\leqslant r\leqslant n$ 。符号 $\binom{n}{r}$（读作"n 选 r"）的定义和表示如下：

$$\binom{n}{r}=\frac{P(n,r)}{P(r,r)}=\frac{n!}{(n-r)!r!}$$

注意：$\binom{n}{0}=1$ ，$\binom{n}{1}=n$ ，$\binom{n}{n}=1$ 。

因为 $n-(n-r)=r$，所以 $\binom{n}{r}=\binom{n}{n-r}$。

与排列数不同，组合数用于不看重顺序的情况。假设我们有 n 个不同的元素，且选取元素的顺序无关紧要，那么从中取出 m 个元素的方法数为 $\binom{n}{m}$，其中 $1\leqslant m\leqslant n$ 。

Example 5

In a group of 5 cats, 5 dogs, and 5 pigs in how many ways can we choose a committee of 5 animals?

例题 5

从 5 只猫、5 只狗和 5 只猪中选择 5 只动物组成一个委员会，共有多少种方法?

(A) 1001 (B) 2002 (C) 3003 (D) 4004 (E) 5005

Solution

Since the order of the chosen animals doesn't matter here, the number of ways is the number of the combinations:

$$\binom{5+5+5}{5}=\frac{15!}{5!\times(15-5)!}=3003$$

The answer should be (C).

解 析

由于选择动物的顺序不重要，选择的方法数应该用组合数来计算：

$$\binom{5+5+5}{5}=\frac{15!}{5!\times(15-5)!}=3003$$

所以答案是 (C)。

13.1.4 The Sum Rule 加法原理

If an event E_1 can happen in n_1 ways, event E_2 can happen in n_2 ways,$\cdots$, event E_k can happen in n_k ways, and if any event E_1, E_2,$\cdots$ or E_k happens, the job is done, then the total ways to do the job is $n_1+n_2+\cdots+n_k$.

假设任务 E_1 有 n_1 种方式完成，任务 E_2 有 n_2 种方式完成，以此类推，任务 E_k 有 n_k 种方式完成，如果 E_1，E_2，$\cdots$，E_k 中的任何一项任务完成，工作就算完成，那么完成该工作的总的方法数是 $n_1+n_2+\cdots+n_k$.

13.1.5 The Product Rule (Fundamental Counting Principle) 乘法原理(基本计数原理)

When a task consists of k separate parts, if the first part can be done in n_1 ways, the second part can be done in n_2 ways, and so on through the kth part, which can be done in n_k ways, then total number of possible results for completing the task is given by the product:

$$n_1\times n_2\times n_3\times\cdots\times n_k$$

假设一项任务由 k 个独立部分组成，如果第一个部分有 n_1 种完成方式，第二个部分有 n_2 种完成方式，以此类推，第 k 个部分有 n_k 种完成方式，那么完成任务的总的方法数由以下乘积给出：

$$n_1\times n_2\times n_3\times\cdots\times n_k$$

Example 6

A girl has 6 shirts, 5 skirts, and 4 pairs of shoes. How many different outfits can she create?

例题 6

一个女孩有 6 件衬衫、5 条裙子和 4 双鞋子，她能搭配出多少种不同的穿着？

(A) 120　　(B) 60　　(C) 15　　(D) 12　　(E) 20

Solution

To make an outfit, the girl needs to determine the shirt, skirt, and shoes. There are 6 choices for shirts, 5 choices for skirts, and 4 choices for shoes. According to the Product Rule, the number of ways is $6\times5\times4=120$. Thus the answer is (A).

解　析

女孩需要决定穿哪件衬衫、哪条裙子和哪双鞋子才能完成一套搭配。衬衫有 6 种选择，裙子有 5 种选择，鞋子有 4 种选择。根据乘法原理，穿搭的种类数为 $6\times5\times4=120$，因此答案是 (A)。

13.1.6 Counting Using Charts 使用图表计数

Example 7

When three positive integers a, b, and c are multiplied together, their product is 36. Suppose $a<b<c$ In how many ways can the numbers be chosen?

例题 7

3 个正整数 a，b，c 的乘积为 36。假设 $a<b<c$，一共有多少种方式选择这些数？

(A) 0　　(B) 1　　(C) 2　　(D) 3　　(E) 4

Solution

The positive divisors of 36 are:

$$1, 2, 3, 4, 6, 9, 12, 18, 36.$$

Since $a<b<c$, we know $b \leqslant c \leqslant 18$, so we apply casework to c:

- If $c=18$, then $(a,b,c)=(1,2,18)$.
- If $c=12$, then $(a,b,c)=(1,3,12)$.
- If $c=9$, then $(a,b,c)=(1,4,9)$.
- If $c=6$, then $(a,b,c)=(2,3,6)$.

Thus, there are 4 ways to choose a, b, and c. The answer is (E).

解 析

36的正整数因子是

$$1，2，3，4，6，9，12，18，36$$

因为$a<b<c$，且$b \leqslant c \leqslant 18$，对$c$进行分类讨论可得：

- 如果$c=18$，那么$(a, b, c)=(1, 2, 18)$。
- 如果$c=12$，那么$(a, b, c)=(1, 3, 12)$。
- 如果$c=9$，那么$(a, b, c)=(1, 4, 9)$。
- 如果$c=6$，那么$(a, b, c)=(2, 3, 6)$。

总共可以有4种选择方式。答案是(E)。

Example 8

Carlos has a set of cards numbered from 1 through 9. He wants to organize these cards into three different sets, each containing three cards. The goal is for the sum of the numbers in each set to be identical. How many ways can Carlos arrange these cards into the desired sets?

例题 8

卡洛斯有一套编号1～9的卡片。他想将这些卡片分成3组，每组包含3张卡片，并且使每组中的数字之和相同。问卡洛斯可以有多少种不同的方式将这些卡片分组?

(A) 0　　(B) 1　　(C) 2　　(D) 3　　(E) 4

Solution

The sum of the nine numbers is 45. Since the sums of numbers in each group are the same, we know it should be 15. Now we need to decompose 15 as the sum of three numbers of 1, 2, ⋯, 9. Before applying the casework, we observe that no two of the cards with numbers 9, 8, and 7 can be in the same group, otherwise the sum of three cards will be greater than 15.

For the group with 9, we can choose the other two numbers as 1 and 5. Then the group with 8 should contain 3 and 4, while the group with 7 contains 2 and 6. If we choose the other two numbers in the group with 9 to be 2 and 4, then the group with 8 must contain 1 and 6, and the group with 7 should include 3 and 5. And there is no other way to choose the other two numbers for the group with 9, so the answer should be (C).

解　析

已知 1 ～ 9 的和是 45，且每组数的和相同，因此每组的和应为 15。即将 1 ～ 9 进行分组，每组和为 15。注意：7，8，9 中的任意两个都不能放在同一组中，否则该组的 3 个数之和会超过 15。

如果我们选择将数字 1 和 5 放在数字 9 的那一组，那么数字 8 那一组必须包含 3 和 4，数字 7 那一组必须包含 2 和 6。如果我们将 2 和 4 放在数字 9 的那一组，则数字 8 那一组包含 1 和 6，而数字 7 的那一组则包含 3 和 5。除此以外，没有其他可能，因此答案是 (C)。

13.1.7 Counting Using Tree Diagram 用树状图计数

You can use a tree diagram to list all the possible outcomes of events.

使用树状图，列出事件的所有可能结果。

Example 9

A designer has 4 fabric colors he may use for a dress: red, green, yellow, and blue. 5 different patterns are available for the dress. If each dress design requires one color and one pattern, how many different dress designs are possible?

例题 9

一位设计师使用 4 种颜色设计一条连衣裙：红色、绿色、黄色和蓝色。另外他还有 5 种不同的款式可供选择。如果每款连衣裙的设计需要一种颜色和一种款式，那么一共能有多少种不同的设计方式？

(A) 10　　(B) 24　　(C) 12　　(D) 14　　(E) 20

Solution

We can draw a tree diagram whose nodes represent the colors of a dress. Below each color are the nodes for patterns. Then, we can count there are $4\times5=20$ designs. Then answer is (E).

解　析

根据题意，绘制树状图来进行计算，首先让每个节点代表一种颜色，然后在每个颜色节点以下画 5 个节点代表不同的款式。这样就一共有 $4\times5=20$ 种不同的设计方式。答案是 (E)。

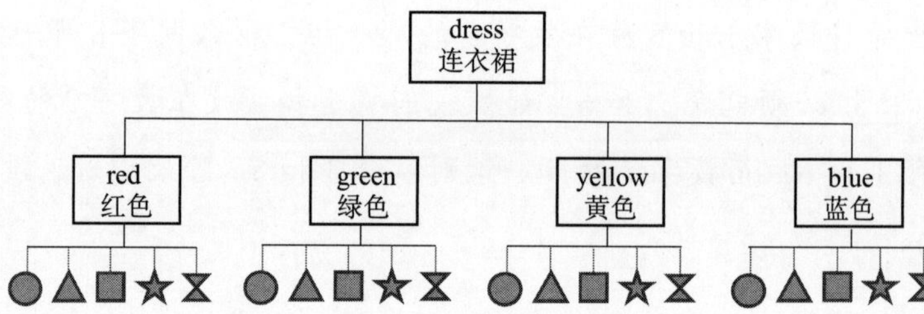

13.2 Circular Permutations 循环排列

THEOREM 1　The number of circular permutations (arrangements in a circle) of n distinct objects is $(n-1)!$.

We can think of this as n people being seated at a round table. Since a rotation of the table does not change an arrangement, we can put person A in one fixed place and then consider the number of ways to seat all the others. Person B can be treated as the first person to seat and M the last person to seat. The number of ways to arrange persons A to M on a circle is the same as the number of ways to arrange persons B to M in a row. So the number of ways is $(n-1)!$.

定理 1 n 个不同物体的循环排列（排成一个圆）数是 $(n-1)!$。

可以将此方式理解为 n 个人坐在一张圆桌旁。由于旋转桌子不会改变排列方式，因此可将 A 固定在一个位置，然后计算安排其他人的方式。B 可以被看成是第一个安排坐下的人，而 M 可以被看成是最后一个被安排落座的人。所以把 A 到 M 安排在圆桌周围的方式数等于将 B 到 M 安排在同一排的方式数。因此，排列的数量是 $(n-1)!$。

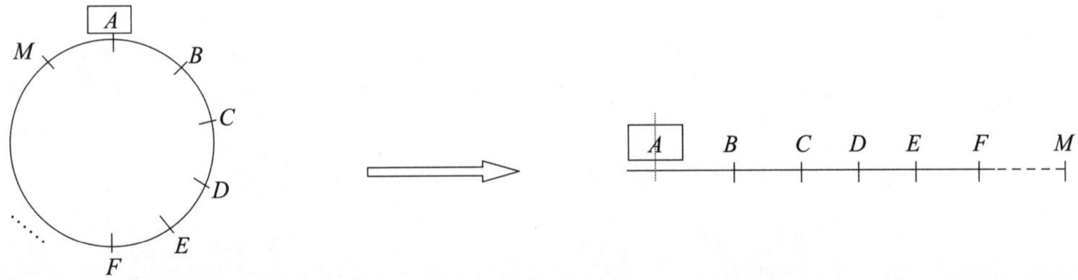

Example 10

In how many ways is it possible to seat eight people at a round table if Alex and Bob must not sit in adjacent seats?

例题 10

将 8 个人安排在一张圆桌上并使亚历克斯和鲍勃不相邻的方式有多少种？

(A) 5040 (B) 3600 (C) 1200 (D) 720 (E) 120

Solution

The number of ways to seat eight people at a round table is 7!=5040. To make Alex and Bob set together, we can bind them and permute other six with them, as if there were 7 people. Then number of such permutation is $2\times 6!=1440$. Thus, there are $5040-1440=3600$ ways to have Alex and Bob apart from each other. The answer is (B).

解 析

将 8 个人安排在一张圆桌上的方法有 7!=5040 种。根据题目，可假设亚历克斯和鲍勃绑定，并和剩下的 6 人排列，共计 7 人，可得 $2\times 6!=1440$ 种排序。因此，亚历克斯和鲍勃分开坐的方式有 $5040-1440=3600$ 种。答案是 (B)。

13.3 Combinations with Repetitions 可重复组合

THEOREM 2 Let n be a positive integer. The number of positive integer solutions to $x_1+x_2+\cdots+x_r=n$ is $\binom{n-1}{r-1}$.

定理 2 设 n 为正整数。方程 $x_1+x_2+\cdots+x_r=n$ 的正整数解的个数为 $\binom{n-1}{r-1}$。

Example 11

Four friends have a total of 9 identical pencils, and each one has at least one pencil. In how many ways can this happen?

例题 11

4 个朋友共有 9 支相同的铅笔，且每个人至少有一支铅笔。一共有多少种可能的情况？

(A) 6　　(B) 24　　(C) 56　　(D) 126　　(E) 3024

Solution

This is equivalent to find the number of positive solutions to the equation $x_1+x_2+x_3+x_4=9$. According to the above theorem, the number is $\binom{8}{3}=56$. The answer is (C).

解　析

这个问题等价于寻找方程 $x_1+x_2+x_3+x_4=9$ 的正整数解。根据上面的定理，可得 $\binom{8}{3}=56$。答案是 (C)。

THEOREM 3 Counting How Many Rectangles

For a rectangular grid with m vertical lines and n horizontal lines, the total number of rectangles that can be counted is: $N=\binom{m}{2}\times\binom{n}{2}$.

定理 3 计算矩形的数量

在一个有 m 条垂直线和 n 条水平线的矩形网格中矩形的总数为 $N=\binom{m}{2}\times\binom{n}{2}$。

THEOREM 4 Rising (Increasing) Number

A rising number, such as 34689, is a positive integer where each digit is larger than the one to its left. The number of integers with n digits in increasing order can be calculated by $\binom{9}{n}$.

定理 4 递增数

递增数，如 34689，是一个每个数字都比它左边的数字大的正整数。n 位递增数的数量可以通过以下公式计算：$\binom{9}{n}$。

THEOREM 5 Falling (Decreasing) Number

A falling number is an integer whose decimal representation has the property that each digit except the unit's digit is larger than the one to its right. For example, 96521 is a falling number but 89642 is not. The number of integers with n digits in decreasing order can be calculated: $\binom{10}{n}$.

定理 5 递减数

递减数是一个除个位数字外，每个数字都比其右边的数字大的正整数。例如，96521 是一个递减数，但 89642 不是。n 位递减数的数量可以通过以下公式计算：$\binom{10}{n}$。

Example 12

How many integers between 2050 and 2500 have four distinct digits arranged in increasing order? (For example, 2347 is one integer.)

例题 12

在 2050 和 2500 之间存在多少个 4 位递增数？（例如 2347 就是这样一个整数。）

(A) 10 (B) 15 (C) 25 (D) 72 (E) 84

Solution

Since both 2050 and 2500 have 2 as the thousands, the hundreds of the increasing number must be greater or equal to 3. So, it suffices for us to find the number of increasing numbers between 300 and 500. The number of increasing numbers between 300 and 400 is $\binom{6}{2}=15$, and the number of increasing number between 400 and 500 is $\binom{5}{2}=10$. Thus the total number of increasing number is 25. The answer is (C).

解　析

由于 2050 和 2500 的千位数字都是 2，因此递增数的百位必须大于等于 3。因此，只要找到 300 和 500 之间的递增数的数量即可。在 300 到 400 之间有 $\binom{6}{2}=15$ 个递增数，而在 400 到 500 之间有 $\binom{5}{2}=10$ 个递增数。因此，共有 25 个递增数，答案是 (C)。

PRACTICE 练习

Problem 1

The Pineview Elementary School debate team consists of two junior members and three senior members. The team coach wants to arrange them for a team photo to be used on the school's website. He plans to have the juniors stand at each end with the three seniors in between. How many ways can the team be arranged for the photo?

问题 1

某小学辩论队由 2 名初级成员和 3 名高级成员组成。教练想为辩论队拍摄团队照片放在学校网站上，他计划让 2 名初级成员站在两端，3 名高级成员站在中间。问这样的排列有多少种不同的方式?

(A) 2 (B) 4 (C) 5 (D) 6 (E) 12

Problem 2

A garden has six gates. In how many ways can Daisy the Dog enter the garden through one gate and exit through a different gate?

问题 2

一个花园有 6 扇门，狗狗黛西有多少种方式从一个门进入花园，又从另一个门离开花园？

(A) 12 (B) 15 (C) 18 (D) 30 (E) 36

Problem 3

Coach Thompson's middle school soccer club is preparing for the upcoming tri-match event. The club has four promising players: Alex, Jamie, Kim, and Max. How many ways can Coach Thompson select three players to start the match?

问题 3

汤普森教练的中学足球俱乐部正在为即将到来的三人足球赛做准备。俱乐部有 4 名球员：亚历克斯、杰米、金姆和马克斯。汤普森教练可以用多少种不同的方式选择 3 名球员开始比赛？

(A) 2 (B) 4 (C) 6 (D) 8 (E) 10

Problem 4

How many ways can 6 distinct vinyl records and 1 special edition CD be arranged in a rack if the special edition CD must be positioned at the far-right end of the rack?

问题 4

有多少种方法可以将 6 张黑胶唱片和 1 张限定版 CD 放在架子上，并使 CD 必须放在架子的最右边？

(A) 7 (B) 24 (C) 90 (D) 120 (E) 720

Problem 5

Tom wants to walk from his home to the park, which is located three blocks west and two blocks south of his house. After walking each block, Tom can choose to continue either west or south, but he needs to steer clear of a risky corner that is one block west and one block south of his home. In how many ways can Tom reach the park by walking a total of five blocks?

问题 5

汤姆想从家里走到公园，公园位于他家以西三个街区、以南两个街区。每走过一个街区后，汤姆可以选择继续向西走或向南走，但他需要避开位于家西边一个街区，再以南一个街区的危险路口。如果汤姆想走过 5 个街区到达公园，一共有多少种方式？

(A) 4　　(B) 5　　(C) 6　　(D) 8　　(E) 10

Problem 6

How many integers between 100 and 999 have three distinct digits?

问题 6

在 100 到 999 之间的整数中，有多少个数包含 3 个不同的数字？

(A) 432　　(B) 648　　(C) 720　　(D) 925　　(E) 950

Problem 7

Chef Gordon has nine different spice jars lined up in his kitchen: two for curry, three for herbs, and four for peppers. How many ways can he arrange the nine jars on a shelf while keeping the curry jars together and keeping the pepper jars together?

问题 7

戈登厨师在他的厨房里排列了 9 种不同的香料罐：2 种咖喱，3 种草药，4 种辣椒。他有多少种排列这 9 个罐子的方法，使得 2 个咖喱罐放在一起，以及 4 个辣椒罐子放在一起？

(A) 1440　　(B) 2880　　(C) 5760　　(D) 182440　　(E) 362880

Problem 8

Eric, Fiona, George, and Hannah are planning to take a road trip in a car that has 4 seats: 1 driver's seat, 1 front passenger seat, and 2 rear passenger seats. Only Fiona and George have valid driver's licenses. How many different seating arrangements can they make for their trip?

问题 8

埃里克，菲奥娜，乔治和汉娜计划开一辆 4 座的汽车去旅行，这辆汽车有 1 个驾驶员座位、1 个前排乘客座位和 2 个后排乘客座位。假设只有菲奥娜和乔治有驾照，他们有多少种不同的座位安排方式？

(A) 2 (B) 4 (C) 6 (D) 12 (E) 24

Problem 9

In the Grand Chess Tournament, a regional chess competition, each participant plays against every other participant exactly once. If a total of 21 matches were played during the event, how many players were competing in the tournament?

问题 9

在一场区域性国际象棋比赛中，每位参赛者都需要与其他参赛者对弈一次。如果在比赛中总共进行了 21 场对弈，那么有多少名棋手参加了这次比赛?

(A) 6 (B) 7 (C) 8 (D) 9 (E) 10

Problem 10

How many 3-digit positive integers have digits whose product equals 18?

问题 10

有多少个 3 位正整数的数字乘积等于 18？

(A) 12 (B) 15 (C) 18 (D) 21 (E) 24

Problem 11

How many different four-digit numbers can be formed by rearranging the four digits in 2009?

问题 11

通过重新排列 2009 中的 4 个数字，可以得到多少个不同的四位数?

(A) 4 (B) 6 (C) 16 (D) 24 (E) 81

Problem 12

How many 4-digit numbers greater than 1000 are there that use the four digits of 2024?

问题 12

使用 2024 中的 4 个数字，可以形成多少个大于 1000 的四位数?

(A) 6 (B) 7 (C) 8 (D) 9 (E) 12

Problem 13

Marcus tosses two regular six-sided dice during a board game. He observes that the product of the numbers showing on the dice is divisible by 6. Which of the following integers cannot be the total of the numbers on the two dice?

问题 13

马库斯在玩桌游时掷了两个标准的六面骰子，他注意到两个骰子的数字之积可以被 6 整除。以下哪个整数不可能是这两个骰子的数字的和？

(A) 5 (B) 6 (C) 7 (D) 8 (E) 9

Problem 14

In how many ways can a family of seven people be seated at a round table if the youngest kid must sit between the parents?

问题 14

如果一个七口之家中最小的孩子必须坐在父母中间，有多少种方法可以将他们安排在一张圆桌上？

(A) 12 (B) 24 (C) 32 (D) 36 (E) 48

Problem 15

Laura is ordering a meal at a casual dining restaurant where she selects one type of protein, two different sides, and one drink. If the order of selection doesn't matter, how many different meal combinations could she choose?

- Protein options: chicken, beef, fish
- Side options: fries, salad, mashed potatoes, coleslaw
- Drink options: soda, lemonade, tea, water

问题 15

劳拉在一家餐厅点餐，她要选择一种肉类，两种不同的配菜和一种饮料。如果选择的顺序无关紧要，她能有多少种不同的餐饮组合？

- 肉类：鸡肉、牛肉、鱼
- 配菜：薯条、沙拉、土豆泥、卷心菜沙拉
- 饮料：苏打水、柠檬水、茶、水

(A) 4 (B) 24 (C) 72 (D) 80 (E) 144

Problem 16

Sticks are used to create a rectangular grid that is 60 sticks long and 32 sticks wide. How many sticks are required in total to complete the grid?

问题 16

建造一个长用 60 根木棍、宽用 32 根木棍的网格围栏，总共需要多少根木棍？

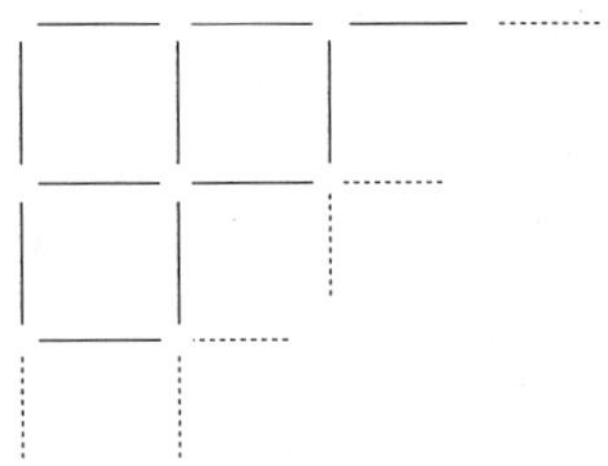

(A) 1920 (B) 1952 (C) 1980 (D) 2013 (E) 3932

Problem 17

Lucas has four distinct toy cars: a Racer, a Speedster, a Cruiser, and a Rocket. He wants to line them up on his windowsill, but he does not want the Cruiser and the Rocket to be placed next to each other. In how many ways can he arrange these toy cars?

问题 17

卢卡斯有 4 辆不同的玩具车：赛车、跑车、警车和火箭车。他想把它们排列在窗台上，但不希望警车和火箭车相邻。他可以有多少种方式排列这些玩具车？

(A) 6 (B) 8 (C) 12 (D) 18 (E) 24

Problem 18

How many 4-digit positive integers have four different digits, which is a multiple of 5, and whose leading digit is not zero and 7 is the largest digit?

问题 18

一个四位正整数由 4 个不同的数字组成，它是 5 的倍数，首位不是 0，并且 7 是其中最大的数字。这样的四位数一共有多少个？

(A) 24 (B) 48 (C) 120 (D) 170 (E) 180

Problem 19

How many positive integers less than 999 contain exactly one 0?

问题 19

有多少小于 999 的正整数恰好包含一个 0？

(A) 81 (B) 99 (C) 153 (D) 171 (E) 189

Problem 20

A coach has five distinct medals to award to three athletes. Each athlete must receive at least one medal. How many ways can the medals be distributed among the athletes?

问题 20

一位教练有 5 枚不同的奖牌要颁给 3 名运动员，已知每名运动员至少获得一枚奖牌。问这些奖牌有多少种不同的分配方式？

(A) 120 (B) 150 (C) 180 (D) 210 (E) 240

Chapter 14 Probability

第 14 章 概率

14.1 Basic Knowledge 基础内容

14.1.1 Properties of Probability 概率的性质

(1) The probability of an event is between 0 and 1.

(2) The probability of an impossible event is 0.

(3) The probability of a certain event is 1.

(4) The probability that an event will occur is equal to one minus the probability that it will not occur.

（1）一个事件的概率大小在 0 和 1 之间。

（2）不可能事件的概率是 0。

（3）必然事件的概率是 1。

（4）某事件发生的概率等于 1 减去该事件不发生的概率。

14.1.2 Basic Probability Formula 基本概率公式

$$\text{Probability}=\frac{\text{number of ways that a certain outcome can occur}}{\text{total number of possible outcomes}}$$

$$\text{概率}=\frac{\text{特定事件的数量}}{\text{所有事件的数量}}$$

Example 1

A bag contains 5 red chips, 3 blue chips, and 4 white chips. If one chip is drawn at random, what is the probability that the chip will not be red?

例题 1

一个袋子里有 5 个红色筹码、3 个蓝色筹码和 4 个白色筹码。如果随机抽取一个筹码，抽到非红色筹码的概率是多少？

(A) $\frac{5}{12}$ (B) $\frac{7}{12}$ (C) $\frac{3}{4}$ (D) $\frac{5}{9}$ (E) $\frac{7}{9}$

Solution

The total number of chips is 12, and the number of non-red chips is 7. Thus, the answer should be (B) $\frac{7}{12}$.

解 析

袋子里一共有 12 个筹码，其中非红色筹码的数量是 7，因此，抽中非红色筹码的概率是 $\frac{7}{12}$，故选 (B)。

14.1.3 Basic Geometric Probability Formula 基本几何概率公式

"Geometric probability" is the same as basic probability, except that we are dealing with the geometric figures instead of the "numbers".

The basic probability formula becomes:

$$P=\frac{\text{measure of geometric figure representing desired outcomes in the event}}{\text{measure of geometric figure representing all outcomes in the same space}}$$

Some examples of geometric measures are lengths, areas, angle measures, and volumes.

For example, a probability determined by comparing the area of a given section to that of a total available region, is:

$$P=\frac{\text{measure of area of favorable region}}{\text{measure of area of total region}}$$

"几何概率"与一般概率的计算是一样的，只不过我们处理的是几何图形而不是"数字"。

概率公式变为

$$P=\frac{\text{代表特定结果的几何度量}}{\text{代表所有结果的几何度量}}$$

几何度量的例子包括长度、面积、角度和体积等。

例如，通过比较给定区域面积与总面积来确定的概率为

$$P=\frac{\text{给定区域面积}}{\text{总面积}}$$

Example 2

If a dart hits the square board ($a = 20$) below, what is the probability that it will land in the circle ($r = 8$)?

例题 2

如果一枚飞镖击中了下面的正方形靶（边长 a =20），那么它落在圆内（半径 $r = 8$）的概率是多少？

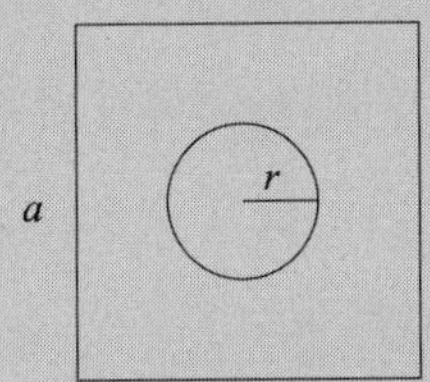

(A) $\frac{2\pi}{25}$　(B) $\frac{\pi}{16}$　(C) $\frac{4\pi}{25}$　(D) $\frac{3\pi}{16}$　(E) $\frac{3\pi}{8}$

Solution

The area of the square is $a^2 = 20^2 = 400$, and the area of the circle is $\pi r^2 = 64\pi$. Therefore, the probability is $\frac{64\pi}{400}=\frac{4\pi}{25}$. The answer is (C).

解　析

正方形的面积是 $a^2 = 20^2 = 400$，圆的面积是 $\pi r^2 = 64\pi$。因此，概率为 $\frac{64\pi}{400}=\frac{4\pi}{25}$。答案是 (C)。

14.1.4 General Addition Rule of Probability 概率的加法规则

(1) If A and B are any two events:

$$P(A\,or\,B)=P(A\cup B)=P(A)+P(B)-P(A\cap B)$$

(2) If A and B are mutually exclusive ($P(A\cap B)=0$):

$$P(A\,or\,B)=P(A\cup B)=P(A)+P(B)$$

Note: Any two events that cannot both occur at the same time are called mutually exclusive.

(3) General addition rule of probability (three events):

$$P(A\cup B\cup C)=P(A)+P(B)+P(C)-P(A\cap B)-P(B\cap C)-P(A\cap C)+P(A\cap B\cap C)$$

(1) A 和 B 是任意两个事件

事件 A 或者事件 B 发生的概率 = 事件 A 发生的概率 + 事件 B 发生的概率 − A 和 B 同时发生的概率。

(2) 事件 A 和 B 互斥，即 A 和 B 同时发生的概率为 0

事件 A 或者事件 B 发生的概率 = 事件 A 发生的概率 + 事件 B 发生的概率。

注意：任何两个不能同时发生的事件称为互斥事件。

(3) 概率的加法规则(三个事件)

事件 A 或事件 B 或事件 C 发生的概率 = 事件 A 发生的概率 + 事件 B 发生的概率 + 事件 C 发生的概率 − A 和 B 同时发生的概率 − B 和 C 同时发生的概率 − A 和 C 同时发生的概率 + A 和 B 和 C 同时发生的概率。

Example 3

Two standard dice are rolled and their face values multiplied. What is the probability that the product is prime or ends in 6?

例题 3

掷 2 颗均匀的骰子并将它们的点数相乘，乘积是质数或以 6 结尾的概率是多少？

(A) $\frac{1}{3}$ (B) $\frac{1}{6}$ (C) $\frac{2}{3}$ (D) $\frac{1}{9}$ (E) $\frac{2}{9}$

Solution

To make the product prime, the only way is to have one dice as 1 and the other as 2, 3, or 5. So the probability of prime products is $\frac{3\times 2}{36}=\frac{1}{6}$. To make the product end in 6, the only way is to have one dice as 1 and the other as 6, or one as 2 and the other as 3, or two dices as 4, or two dices as 6. So the probability of products ending in 0 is $\frac{2\times 2+2}{36}=\frac{1}{6}$. Note that these two events are exclusive. Thus, the answer should be $\frac{1}{6}+\frac{1}{6}=\frac{1}{3}$. Then answer is (A).

解 析

使乘积为质数的方法是一个骰子的点数为 1，另一个骰子的点数为 2、3 或 5。因此，乘积为质数的概率是 $\frac{3\times 2}{36}=\frac{1}{6}$。使乘积以 6 结尾的方法是一个骰子的点数为 1，另一个骰子的点数为 6，或者一个点数为 2，另一个为 3，或者两个骰子点数为 4，或两个骰子点数为 6。因此，乘积以 6 结尾的概率是 $\frac{2\times 2+2}{36}=\frac{1}{6}$。由于这两种情况是互斥的。答案应为 $\frac{1}{6}+\frac{1}{6}=\frac{1}{3}$。答案是 (A)。

14.1.5 Two Way Tables 双向表

Two-way tables are used to display the frequency information of two categories of variables collected from a common source. The table has labels for the categories at the top and on the left, and the frequency data is shown in four or more of the cells in the table. The "total" of each row can be seen at the right and "total" of each column at the bottom. This table is also known as a contingency table. These tables are often used to analyze survey results.

Here is an example of a two-way table that shows the number of males and females who like various games:

	Baseball	Basketball	Football	Total
Male	13	15	20	48
Female	23	16	13	52
Total	36	31	33	100

Each entry in the table is called joint frequency and the sum or total of rows and columns is called the marginal frequency. Here, 13, 15, 23, $\cdots$, are joint frequencies, and the sums, 48, 52, 36, 31,$\cdots$, are the marginal frequencies.

双向表用于显示从同一来源处收集的两个变量的频率信息。表格的顶部和左侧是类别标签，频率数据显示在表格中的 4 个或更多单元格中。每行的“总数”显示在右侧，每列的“总数”显示在底部。该表也称为列联表，常用于分析调查结果。

以下是双向表的一个范例，统计了喜欢各种游戏的男性和女性的人数：

	棒球	篮球	足球	总计
男性	13	15	20	48
女性	23	16	13	52
总计	36	31	33	100

表中的每个条目称为联合频率，每行和每列的和称为边际频率。在这里，13、15、23 等是联合频率，而 48、52、36、31 等和则是边际频率。

Example 4

Two dices are thrown. What is the probability that the product of the two numbers is a multiple of 4?

例题 4

掷出 2 个骰子，问 2 个数字的乘积是 4 的倍数的概率是多少？

(A) $\frac{1}{9}$　(B) $\frac{1}{6}$　(C) $\frac{1}{3}$　(D) $\frac{13}{36}$　(E) $\frac{4}{9}$

Solution

We can use a two-way table to solve this problem.

D_2 \ D_1	1	2	3	4	5	6
1	1	2	3	4	5	6
2	2	4	6	8	10	12
3	3	6	9	12	15	18
4	4	8	12	16	20	24
5	5	10	15	20	25	30
6	6	12	18	24	30	36

List the numbers of multiple of 4: 4, 4, 4, 8, 8, 12, 12, 16, 20, 20, 24, 24, 36. The probability that the product is multiple of 4 is: $\frac{13}{36}$. The answer is (D).

解 析

使用双向表解决这个问题。

D_2 \ D_1	1	2	3	4	5	6
1	1	2	3	4	5	6
2	2	4	6	8	10	12
3	3	6	9	12	15	18
4	4	8	12	16	20	24
5	5	10	15	20	25	30
6	6	12	18	24	30	36

根据表格，4 的倍数有：4、4、4、8、8、12、12、16、20、20、24、24、36。所以乘积是 4 的倍数的概率是 $\frac{13}{36}$。答案是 (D)。

14.1.6 Tree Diagrams 树形图

A probability tree diagram is used to represent the probability of occurrence of events without using complicated formulas. It displays all the possible outcomes of an event.

There are two main parts of a probability tree. These are the nodes and the branches. The nodes can further be classified into a parent node and a sibling node. The parent node represents a certain event and has a probability of 1. The sibling nodes denote other additional possible events or outcomes. The branches denote the probability of occurrence of these events.

Suppose a fair coin is tossed once, then the probability tree can be constructed as follows:

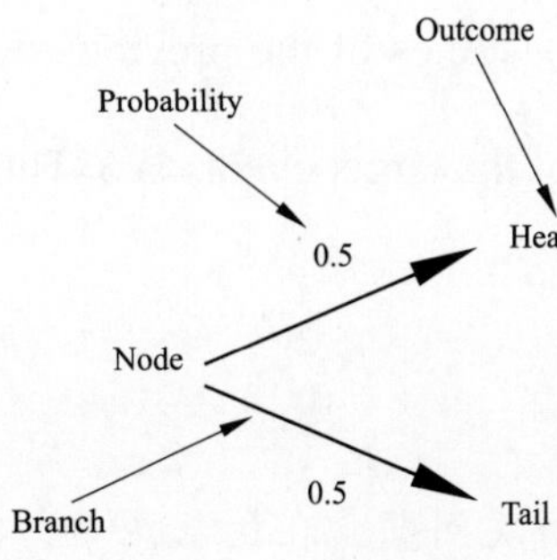

我们可以用概率树形图直观表示事件的所有可能结果，得到事件发生的概率，从而避免使用复杂公式计算。

概率树主要有两个部分。它们是节点和分支。节点可以进一步分为父节点和子节点。父节点表示某个事件，其发生概率为 1，子节点表示该事件发生后可能的结果，而分支表示这些结果发生的概率。

假设有一枚均匀的硬币被抛掷一次，那么其概率树如下图所示：

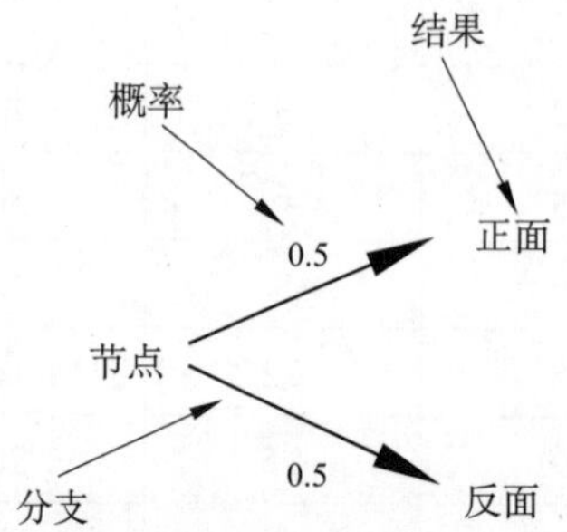

Example 5

If three coins are tossed, what is the probability that exactly two of them show heads or exactly two of them show tails?

例题 5

如果掷 3 枚硬币，那么恰好有 2 枚硬币显示正面或恰好有 2 枚硬币显示反面的概率是多少？

(A) $\frac{1}{8}$　　(B) $\frac{1}{4}$　　(C) $\frac{3}{8}$　　(D) $\frac{1}{2}$　　(E) $\frac{3}{4}$

Solution

We can draw a probability tree diagram by setting the nodes on the nth layer (setting the starting node as layer 0) as the outcomes of the previous n tossing. So, the total number of events is $2^3=8$, and the number of the target events is 6. Therefore, the probability is $\frac{6}{8}=\frac{3}{4}$. The answer is (E).

解　析

可通过画出概率树状图来解决问题，在树状图中，第 n 层的节点（将起始节点设为第0层）为前n次投掷的结果。因此，事件总数是$2^3=8$，而目标事件数是6。因此，概率是$\frac{6}{8}=\frac{3}{4}$。答案是 (E)。

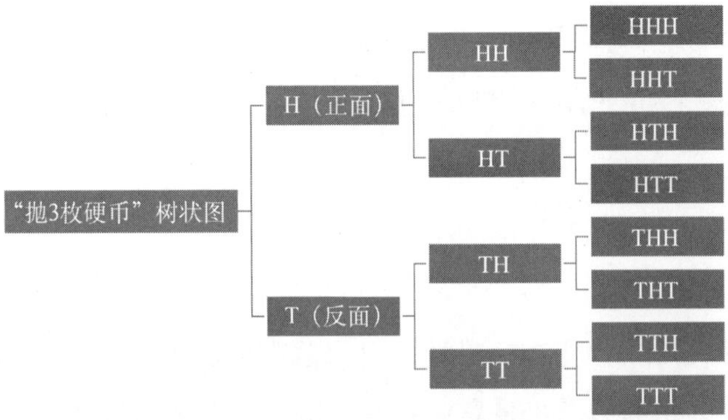

Example 6

A cricket randomly hops between 4 leaves, on each turn hopping to one of the other 3 leaves with equal probability. After 4 hops what is the probability that the cricket has returned to the leaf where it started?

例题 6

一只蟋蟀在 4 片叶子之间随机跳跃，每次跳跃到另外 3 片叶子上的概率相同。经过 4 次跳跃后，蟋蟀回到起始叶子的概率是多少？

(A) $\frac{2}{9}$　　(B) $\frac{19}{80}$　　(C) $\frac{20}{81}$　　(D) $\frac{1}{4}$　　(E) $\frac{7}{27}$

Solution

Before each hop, the cricket has 3 choices for the next stop. So, the total number of outcomes is $3^4=81$. If the last stop of the cricket is the beginning leaf, then the choices of the 3rd stop will give us 2 cases.

If the 3rd stop is the beginning leaf, then the 2nd and 4th stop both have 3 choices, resulting in 9 different ways. If the 3rd stop is not the beginning leaf, then the 2nd and 4th stop have 2 choices, respectively. There are $3\times2\times2=12$ ways. Thus, the probability is $\frac{9+12}{81}=\frac{7}{27}$. The answer is (E).

解　析

由于蟋蟀的每次跳跃都有 3 种选择，因此总事件数为 $3^4=81$。如果最后一次停靠的是起始树叶，那么根据第三次停留的树叶，有以下 2 种情况。

若第三次停靠的是起始树叶，那么第二和第四次停靠的树叶都有 3 种可能，所以有 9 种不同的情况。若第三次停靠的不是起始树叶，则第二次和第四次停靠的地点分别有 2 种可能，一共有 $3\times2\times2=12$ 种情况。因此，蟋蟀回到起始树叶的概率是 $\frac{9+12}{81}=\frac{7}{27}$。答案是 (E)。

Example 7

A movie theater has 4 rows of seats with 3 seats in each row. Eight viewers have already taken their seats randomly. A pair of friends is next to enter the theater. What is the probability that there will be 2 adjacent seats available in the same row for the friends?

例题 7

一个电影院有 4 排座位，每排有 3 个座位。影院内已经有 8 名观众随机坐下。这时，2 个好朋友进入影院。这 2 人可以坐在同一排中相邻的座位上的概率是多少？

(A) $\frac{8}{15}$　(B) $\frac{32}{55}$　(C) $\frac{20}{33}$　(D) $\frac{34}{55}$　(E) $\frac{8}{11}$

Solution

Suppose the viewers are indistinguishable. This question is asking that if 4 empty seats are placed, what is the probability that there are 2 adjacent seats open. We proceed by casework:

Case 1: There is exactly one pair of adjacent open seats. Then the other seat in that row must be occupied. The other two empty seats are distributed across the remaining 3 rows without being adjacent, which is $\binom{9}{2}-6=30$ cases per pair of open seats for $30\times 8=240$ total cases.

Case 2: There is one row of open seats. 4 ways to choose the row and 9 ways to choose the final empty seat for $4\times 9=36$ cases.

Case 3: There are 2 independent pairs of open seats. Choose the 2 rows, then the placement of each pair within each row for $\binom{4}{2}\times 4=24$ cases.

In total, we get $240+36+24=300$ cases in total, for a probability of $\frac{300}{495}=\frac{20}{33}$, where $495=\binom{12}{8}$. The answer is (C).

解 析

假设观众是不可区分的。那么这个问题在问，如果有 4 个空座位，有 2 个相邻座位是空的概率有多大。分类讨论如下：

情况 1：有一对相邻的空座位，并且，该排的另一个座位已经被占。其余两个空座位分布在剩下的 3 排中，且不相邻，每对这样的空座位有$\binom{9}{2}-6=30$ 种情况，共有 $30\times 8=240$ 种情况。

情况 2：有一整排空座位。则有 4 种选择空排的方法，9 种选择最后一个空座位的方法，因此共有 $4\times 9=36$ 种情况。

情况 3：有两对相邻的空座位。选择 2 排座位，然后在每排内选择一对座位，共有$\binom{4}{2}\times 4=24$ 种情况。

因此，总共有 $240+36+24=300$ 种情况，所以概率为 $\frac{300}{495}=\frac{20}{33}$，其中 $495=\binom{12}{8}$。答案为 (C)。

PRACTICE 练习

Problem 1

A bag contains 3 red chips, 4 blue chips, and 5 white chips. If a chip is drawn at random, what is the probability that the chip is red or white?

问题 1

一个袋子里有 3 个红色筹码、4 个蓝色筹码和 5 个白色筹码。如果随机抽取一个筹码，抽到红色或白色筹码的概率是多少？

(A) $\frac{1}{9}$ (B) $\frac{2}{9}$ (C) $\frac{1}{3}$ (D) $\frac{2}{3}$ (E) $\frac{7}{9}$

Problem 2

A classroom has 6 freshmen, 18 sophomores, and 12 juniors. After enrolling more sophomores, the probability that a randomly selected student is a sophomore is now 60%. How many sophomores did the classroom add?

问题 2

一个教室里有 6 名大一学生、18 名大二学生和 12 名大三学生。增加了一些大二学生后，现在随机选出一名学生是大二学生的概率为 60%。问教室增加了多少名大二学生？

(A) 6 (B) 9 (C) 12 (D) 18 (E) 24

Problem 3

30 cards numbered from 1 to 30 are placed in a box. If a card is selected at random, what is the probability that the card is a prime number or a multiple of seven?

问题 3

30 张编号为从 1 到 30 的卡片被放在一个盒子里。如果随机抽取一张卡片，卡片上的数字是质数或 7 的倍数的概率是多少？

(A) $\frac{1}{30}$ (B) $\frac{2}{25}$ (C) $\frac{2}{5}$ (D) $\frac{13}{30}$ (E) $\frac{7}{10}$

Problem 4

A drawer contains three different socks marked with the numbers 1, 3, and 5. A second drawer contains three different socks marked with the numbers 2, 4, and 6. If one sock is selected from each drawer, how many different values are possible for the sum of the numbers on the selected socks?

问题 4

第一个抽屉里有 3 只不同的袜子，分别标有数字 1、3 和 5。第二个抽屉里有 3 只不同的袜子，分别标有数字 2、4 和 6。如果从每个抽屉中选出一只袜子，两只袜子上的数字之和可能有多少种不同的值？

(A) 2 (B) 3 (C) 4 (D) 5 (E) 6

Problem 5

When a fair six-sided die is tossed on a tabletop, the bottom face cannot be seen. What is the probability that the product of the numbers on the five faces that can be seen is divisible by 4?

问题 5

当一颗均匀的六面骰子在桌面上掷出时，底面的数字不可见。那么 5 个可见面的数字乘积可被 4 整除的概率是多少？

(A) $\frac{1}{3}$ (B) $\frac{1}{2}$ (C) $\frac{2}{3}$ (D) $\frac{5}{6}$ (E) 1

Problem 6

Digit d is randomly selected from the set {4, 5, 6, 7}. Without replacement of d, another digit e is selected. What is the probability that the two-digit number de is a multiple of 4?

问题 6

从集合 {4, 5, 6, 7} 中随机选择一个数字 d，不放回 d，再选择另一个数字 e。两位数 de 是 4 的倍数的概率是多少？

(A) $\frac{1}{6}$ (B) $\frac{1}{4}$ (C) $\frac{1}{3}$ (D) $\frac{2}{3}$ (E) $\frac{3}{4}$

Problem 7

A box contains five cards, numbered 1, 4, 7, 11, and 14. Three cards are selected randomly without replacement from the box. What is the probability that 11 is the largest value selected?

问题 7

一个盒子里有 5 张卡片，编号为 1，4，7，11 和 14。从盒子里抽取三张不同的卡片。抽出的最大数是 11 的概率是多少？

(A) $\frac{1}{10}$ (B) $\frac{1}{5}$ (C) $\frac{3}{10}$ (D) $\frac{2}{5}$ (E) $\frac{1}{2}$

Problem 8

Harold tosses a nickel six times. What is the probability that he gets at least as many heads as tails?

问题 8

哈罗德投掷 6 次五美分硬币，正面出现的次数不少于反面的概率是多少？

(A) $\frac{5}{16}$　　(B) $\frac{3}{8}$　　(C) $\frac{1}{2}$　　(D) $\frac{9}{16}$　　(E) $\frac{21}{32}$

Problem 9

Ten cards numbered 1 through 10 are shuffled and placed face down. One card is selected at random, and a six-sided die is rolled. What is the probability that the product of the number on the card and the number on the die will be a perfect square?

问题 9

10 张标有 1 ～ 10 的卡片被洗牌后面朝下放置。随机选择一张卡片，并掷一枚六面骰子。卡片上的数字与骰子上的数字的乘积是完全平方数的概率是多少？

(A) $\frac{1}{9}$　　(B) $\frac{1}{6}$　　(C) $\frac{11}{60}$　　(D) $\frac{1}{5}$　　(E) $\frac{7}{30}$

Problem 10

An integer between 1000 and 9999, inclusive, is chosen at random. What is the probability that it is an even integer whose digits are all distinct?

问题 10

从 1000 ～ 9999（包含两者）之中随机选择一个整数。这个整数是偶数且所有数字都不同的概率是多少？

(A) $\frac{286}{1125}$　　(B) $\frac{287}{1125}$　　(C) $\frac{32}{125}$　　(D) $\frac{289}{1125}$　　(E) $\frac{58}{225}$

Problem 11

Three concentric circles have radii of 1, 2, and 3. If a point is randomly selected from the interior of the largest circle, what is the probability that it is in the region bounded by the two smaller circles?

问题 11

3 个同心圆的半径分别为 1、2 和 3。如果从大圆的内部随机选取一个点，这个点位于由两个较小的圆界定的区域的概率是多少？

(A) $\frac{5}{27}$ (B) $\frac{1}{3}$ (C) $\frac{2}{3}$ (D) $\frac{1}{9}$ (E) $\frac{2}{9}$

Problem 12

How many non-congruent triangles can be formed by choosing three vertices from eight distinct points arranged on a chessboard as below?

问题 12

在如图所示的棋盘上的 8 个点中选取 3 个点为顶点，构成的不同的三角形有多少个？

(A) 5 (B) 6 (C) 7 (D) 8 (E) 9

Problem 13

Emma has 1 blue and 1 red token in her hand. Liam has 1 blue, 1 yellow, and 2 red tokens in his hand. Each randomly selects a token to show the other. What is the probability that the colors match?

问题 13

艾玛手中有 1 个蓝色和 1 个红色的代币，利亚姆手中有 1 个蓝色、1 个黄色和 2 个红色的代币。每个人随机选一个代币给对方看，这两个代币颜色相同的概率是多少？

(A) $\frac{1}{4}$ (B) $\frac{1}{3}$ (C) $\frac{3}{8}$ (D) $\frac{1}{2}$ (E) $\frac{2}{3}$

Problem 14

Eight marbles are placed in a bag: four red marbles numbered 1, 2, 3, and 4, and four blue marbles numbered 1, 2, 3, and 4. Two marbles are drawn at random. A winning pair is either two marbles of the same color or two marbles with the same number. What is the probability of drawing a winning pair?

问题 14

一个袋子里有 8 个弹珠：4 个红色弹珠，编号为 1，2，3，4，4 个蓝色弹珠，编号为 1，2，3，4。随机抽取 2 个弹珠，如果这 2 个弹珠颜色相同或编号相同，就算是一个获胜组合。问抽到获胜组合的概率是多少？

(A) $\frac{2}{7}$　　(B) $\frac{3}{8}$　　(C) $\frac{1}{2}$　　(D) $\frac{4}{7}$　　(E) $\frac{5}{8}$

Problem 15

A top hat contains 4 red chips and 2 green chips. Chips are drawn randomly, one at a time without replacement, until all 4 of the reds are drawn or until both green chips are drawn. What is the probability that the 4 reds are drawn?

问题 15

一顶高帽里有 4 个红色筹码和 2 个绿色筹码。筹码随机抽取，一次抽取一个，不放回，直到抽取到所有 4 个红色筹码或抽取到 2 个绿色筹码为止。先抽取到 4 个红色筹码的概率是多少？

(A) $\frac{3}{10}$　　(B) $\frac{1}{3}$　　(C) $\frac{2}{5}$　　(D) $\frac{3}{5}$　　(E) $\frac{7}{10}$

Problem 16

A singles tournament had six players. Each player played every other player only once, with no ties. If Helen won 3 games, Ines won 3 games, Janet won 3 games, Kendra won 2 games and Lara won 2 games, how many games did Monica (the sixth player) win?

问题 16

一场单打锦标赛有 6 名选手参赛，每名选手与其他选手只对战一次，没有平局。海伦赢了 3 场比赛，伊内斯赢了 3 场比赛，珍妮特赢了 3 场比赛，肯德拉赢了 2 场比赛，劳拉赢了 2 场比赛。问莫妮卡（第六位选手）赢了几场比赛？

(A) 0　　(B) 1　　(C) 2　　(D) 3　　(E) 4

Problem 17

A bag contains four pieces of paper, each labeled with one of the digits “4” “5” “6” or “7” , with no repeats. Three of these pieces are drawn, one at a time without replacement, to construct a three-digit number. What is the probability that the three-digit number is a multiple of 3?

问题 17

一个袋子里有 4 张纸条，分别标有数字 “4” “5” “6” 和 “7”。随机抽取 3 张纸条，组成一个三位数。这个三位数是 3 的倍数的概率是多少？

(A) $\frac{1}{4}$ (B) $\frac{1}{3}$ (C) $\frac{1}{2}$ (D) $\frac{2}{3}$ (E) $\frac{3}{4}$

Problem 18

Four children—Alex, Brooke, Casey, and Drew—are randomly assigned to sit on the four sides of a square sandbox, one child per side. What is the probability that Alex and Casey end up sitting on opposite sides?

问题 18

亚历克斯、布鲁克、凯西、德鲁 4 个孩子被随机分配坐在一个方形沙盘的四边，每边一个孩子。艾历克斯和凯西坐在相对两边的概率是多少？

(A) $\frac{1}{4}$ (B) $\frac{1}{3}$ (C) $\frac{1}{2}$ (D) $\frac{2}{3}$ (E) $\frac{3}{4}$

Problem 19

A large square region is paved with n^2 gray square tiles, each measuring s inches on a side. A border of d inches wide surrounds each tile. The figure below shows the case for $n=3$. When $n=24$, the 576 gray tiles cover 81% of the area of the large square region. What is the ratio $\frac{d}{s}$ for this larger value of n?

问题 19

一个大正方形区域铺设了 n^2 个灰色正方形瓷砖，每个瓷砖的边长为 s 英寸。瓷砖之间有一条 d 英寸宽的边框。下图显示了 $n=3$ 时的情况。当 $n=24$ 时，包含的 576 个灰色瓷砖覆盖了 81% 的区域。这时比值 $\frac{d}{s}$ 是多少？

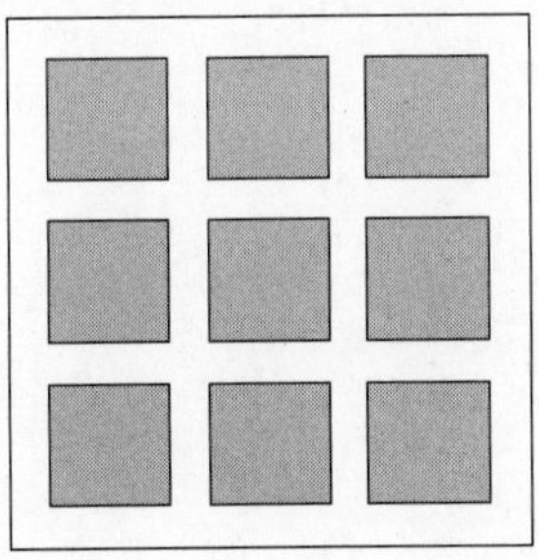

(A) $\frac{24}{225}$ (B) $\frac{6}{25}$ (C) $\frac{1}{4}$ (D) $\frac{9}{25}$ (E) $\frac{9}{16}$

Answers

答案

Chapter 1　第 1 章

Problem　问题 1~10

E　C　B　A　C　C　B　C　C　A

Problem　问题 11~17

B　D　E　A　D　C　C

Chapter 2　第 2 章

Problem　问题 1~10

A　C　D　A　A　B　A　C　E　D

Problem　问题 11~15

E　C　A　A　A

Chapter 3　第 3 章

Problem　问题 1~10

E　D　C　B　B　B　A　C　B　C

Problem 问题 11~16

E　C　A　C　B　C

Chapter 4　第 4 章

Problem　问题 1~10

A　B　A　E　D　C　C　D　D　D

Problem　问题 11~15

B　A　D　E　A

Chapter 5　第 5 章

Problem　问题 1~10

C　D　B　B　E　A　C　C　B　E

Problem　问题 11~20

D　C　A　B　E　E　D　B　B　E

Chapter 6　第 6 章

Problem　问题 1~10

C　A　C　A　A　D　B　B　E　D

Problem　问题 11~20

C　E　E　E　C　A　C　B　C　B

Chapter 7　第 7 章

Problem　问题 1~10

D　D　B　A　A　D　C　A　C　D

Problem　问题 11~18

B　E　D　E　D　B　C　B

Chapter 8　第 8 章

Problem　问题 1~10

A　D　C　D　C　E　B　B　C　C

Problem 问题 11~15

E　C　D　D　D

Chapter 9　第 9 章

Problem　问题 1~10

C　E　E　C　C　C　B　C　A　C

Problem　问题 11~17

B　B　A　E　E　B　B

Chapter 10　第 10 章

Problem　问题 1~10

B　D　E　C　E　B　C　A　A　B

Problem 问题 11~16

B D D B A A

Chapter 11 第 11 章

Problem 问题 1~10

E E B E D E C B E A

Problem 问题 11~15

B A A A A

Chapter 12 第 12 章

Problem 问题 1~10

D C D D D D C C D B

Problem 问题 11~20

D C E C C D C B A E

Chapter 13 第 13 章

Problem 问题 1~10

E D B E A B C D B B

Problem 问题 11~20

B D B E C E C D D B

Chapter 14 第 14 章

Problem 问题 1~10

D B D D E B C E C B

Problem 问题 11~19

B D C D B C C B A

Appendix Formulas

附录 公式

Algebra 代数

1. Basic Operation 基础运算

$$a+b=b+a$$

$$a+(b+c)=(a+b)+c$$

$$a\times b=b\times a$$

$$(a\times b)\times c=a\times(b\times c)$$

$$a\times(b+c)=a\times b+a\times c$$

$$a-(b\pm c)=(a-b)\pm c$$

$$a+(b\pm c)=(a+b)\pm c$$

2. Fraction and Ratio 分数和比

$$\frac{a}{c}+\frac{b}{c}=\frac{a+b}{c}(c\neq 0)$$

$$\frac{a}{c}+\frac{b}{d}=\frac{ad+bc}{cd}(c\neq 0,d\neq 0)$$

$$a\times\frac{b}{c}=\frac{ab}{c}(c\neq 0)$$

$$\frac{a}{c}\times\frac{b}{d}=\frac{ab}{cd}(c\neq 0,d\neq 0)$$

$$\frac{ab}{ac}=\frac{b}{c}(a\neq 0,c\neq 0)$$

$$a+\frac{b}{c}=\frac{ac+b}{c}(c\neq 0)$$

$$\frac{a}{c}\div\frac{b}{d}=\frac{a}{c}\times\frac{d}{b}=\frac{ad}{bc}(b\neq 0,c\neq 0)$$

$$a:b=a\div b=\frac{a}{b}(b\neq 0)$$

$$ac:bc=a:b=\frac{a}{d}:\frac{b}{d}(b\neq 0,c\neq 0,d\neq 0)$$

$$a:b=c:d\Leftrightarrow ad=bc(b\neq 0,d\neq 0)$$

$$\frac{a}{b}=\frac{c}{d}\Leftrightarrow\frac{a+b}{b}=\frac{c+d}{d}\Leftrightarrow\frac{a-b}{b}=\frac{c-d}{d}$$

$$\frac{a}{b}=\frac{c}{d}\Leftrightarrow\frac{a+b}{a-b}=\frac{c+d}{c-d}$$

$$\frac{a}{100}=a\%$$

3. Exponentiation 乘方运算

$$a^m\cdot a^n=a^{m+n}(a\neq 0)$$

$$(a^m)^n=a^{mn}(a\neq 0)$$

$$a^m b^m=(a\cdot b)^m(a\neq 0)$$

$$\frac{a^m}{b^m}=\left(\frac{a}{b}\right)^m$$

4. Radical operation 根式运算

$$a^2=b\Leftrightarrow a=\pm\sqrt{b}(b\geqslant 0)$$

$$\sqrt{a^2}=a(a\geqslant 0)$$

$$\sqrt{a^2}=-a(a\leqslant 0)$$

$$(\sqrt{a})^2=a(a\geqslant 0)$$

$$\sqrt{a^2b}=\sqrt{a^2}\cdot\sqrt{b}=a\sqrt{b}(a\geqslant 0, b\geqslant 0)$$

5. Absolute value operations 绝对值运算

$$|-x|=|x|$$

$$|x-y|=|y-x|$$

$$|xy| = |x| \cdot |y|$$

$$\left|\frac{x}{y}\right| = \frac{|x|}{|y|} \ (y \neq 0)$$

6. Algebraic operations 代数式运算

$$x^2 - y^2 = (x+y)(x-y)$$

$$(x \pm y)^2 = x^2 \pm 2xy + y^2$$

7. Arithmetic sequence 等差数列

$$d = a_{n+1} - a_n$$

$$a_n = a_1 + (n-1)d$$

$$S_n = \frac{(a_1 + a_n)n}{2}$$

$$d = \frac{a_m - a_n}{m - n}$$

8. Geometric Sequences 等比数列

$$q = \frac{a_{n+1}}{a_n}$$

$$a_n = a_1 \cdot q^{n-1} = a_m \cdot q^{n-m}$$

9. Calculation Conclusion 必背计算结论

$2^0 = 1$　　$2^6 = 64$　　$2^{12} = 4096$

$2^1 = 2$　　$2^7 = 128$　　$2^{13} = 8192$

$2^2 = 4$　　$2^8 = 256$　　$2^{14} = 16384$

$2^3 = 8$　　$2^9 = 512$　　$2^{15} = 32768$

$2^4 = 16$　　$2^{10} = 1024$　　$2^{16} = 65536$

$2^5 = 32$　　$2^{11} = 2048$

$11^2 = 121$　　$14^2 = 196$　　$17^2 = 289$

$12^2 = 144$　　$15^2 = 225$　　$18^2 = 324$

$13^2 = 169$　　$16^2 = 256$　　$19^2 = 361$

$1! = 1$ $3! = 6$ $5! = 120$

$2! = 2$ $4! = 24$ $6! = 720$

$\pi \approx 3.1416$ $\sqrt{2} \approx 1.4142$ $\sqrt{5} \approx 2.2361$

$e \approx 2.7183$ $\sqrt{3} \approx 1.7321$ $\sqrt{6} \approx 2.4495$

$3^2 + 4^2 = 5^2$ $7^2 + 24^2 = 25^2$

$5^2 + 12^2 = 13^2$ $8^2 + 15^2 = 17^2$

Geometry 几何

1. Angles and Lines 角和线

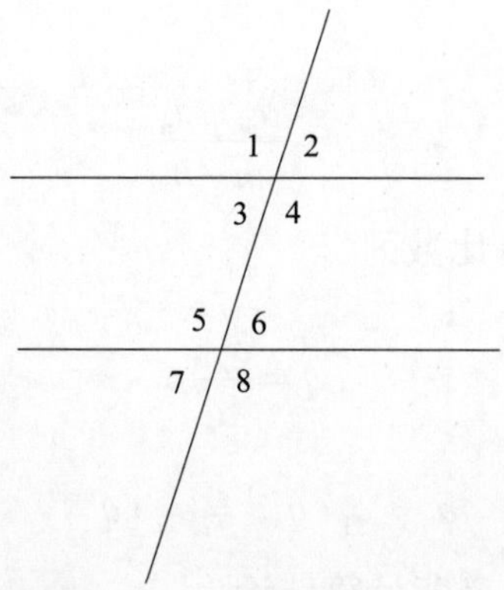

- Sum of complementary angles: $a + b = 90°$
- Sum of supplementary angles: $a + b = 180°$
- 互余的两个角和为 90°
- 互补的两个角和为 180°
- Relationship of Angles Formed by Parallel Lines:

Alternate interior angles: $\angle 3 = \angle 6$, $\angle 4 = \angle 5$

Alternate exterior angles: $\angle 1 = \angle 8$, $\angle 2 = \angle 7$

Corresponding angles: $\angle 1 = \angle 5$, $\angle 2 = \angle 6$, $\angle 3 = \angle 7$, $\angle 4 = \angle 8$

Interior angles on the same sides of transversal: $\angle 3 + \angle 5 = 180°$, $\angle 4 + \angle 6 = 180°$

- 平行线间角的关系（“三线八角”）：

内错角相等：$\angle 3 = \angle 6$, $\angle 4 = \angle 5$

外错角相等：$\angle 1 = \angle 8$, $\angle 2 = \angle 7$

同位角相等：$\angle 1=\angle 5$, $\angle 2=\angle 6$, $\angle 3=\angle 7$, $\angle 4=\angle 8$

同旁内角互补：$\angle 3+\angle 5=180°$, $\angle 4+\angle 6=180°$

- The Property of Triangles:

The area of a triangle = $\frac{1}{2}\times$ base $\times$ height

The sum of any two sides in a triangle is greater than the third side, and the difference between any two sides is less than the third side.

- 三角形的性质

三角形的面积 = $\frac{1}{2}\times$底$\times$高

三角形两边之和大于第三边，两边之差小于第三边。

- Pythagoras Theorem

$$a^2+b^2=c^2$$

In a right triangle, the sum of the squares of the two right-angled sides is equal to the square of the hypotenuse.

($3^2+4^2=5^2$, $5^2+12^2=13^2$, $7^2+24^2=25^2$, $8^2+15^2=17^2$.)

- 勾股定理

$$a^2+b^2=c^2$$

在直角三角形中，两条直角边的平方和等于斜边的平方。

（ $3^2+4^2=5^2$, $5^2+12^2=13^2$, $7^2+24^2=25^2$, $8^2+15^2=17^2$ 。）

- Mid-line of triangle 三角形中线性质

$$BD=DC\ ,\ S_{\triangle ABD}=S_{\triangle ADC}$$

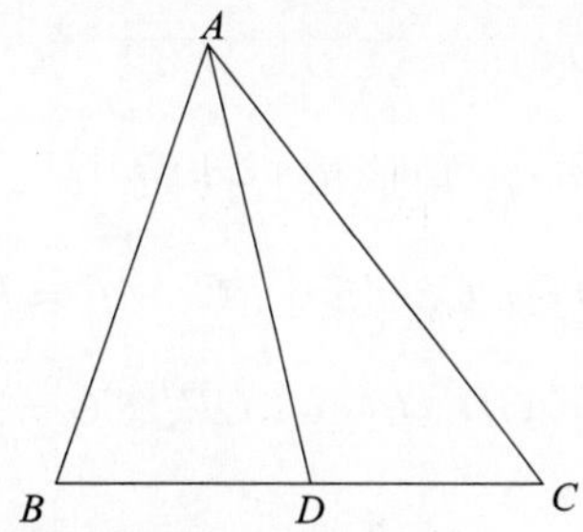

- Triangle bisector 三角形角平分线性质

$$DE=DF\ ,\ \angle BAD=\angle CAD$$

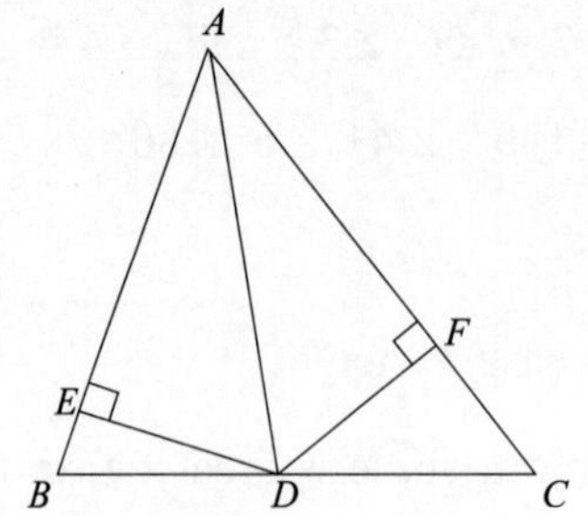

- Altitude of triangle 三角形垂线性质

$$AD \perp BC\ ,\ S_{\triangle ABC} = \frac{1}{2} \times AD \times BC$$

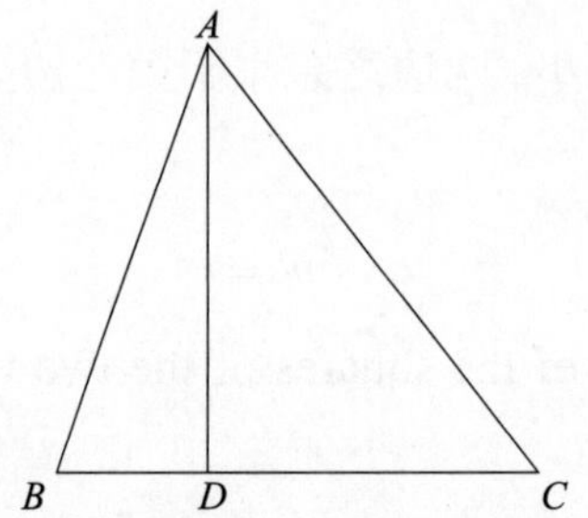

- Vertical bisector of a triangle 三角形垂直平分线性质

$$BD = DC,\ DE \perp BC,\ FB = FC$$

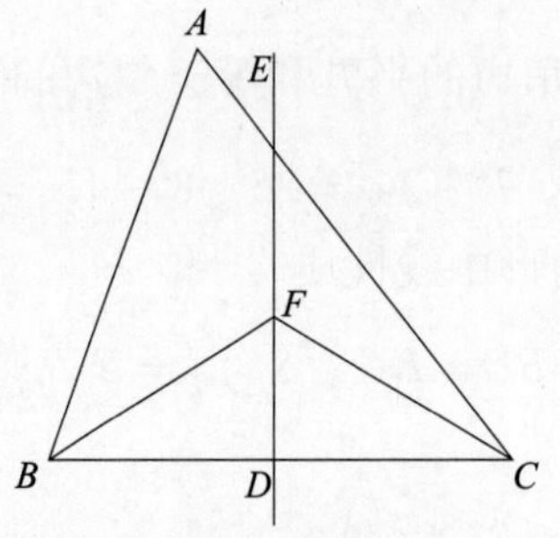

- Center of gravity of triangle 三角形重心性质

$$BD = DC,\ AE = EC,\ AF = FB$$

$$AG = 2GD,\ BG = 2GE,\ CG = 2GF$$

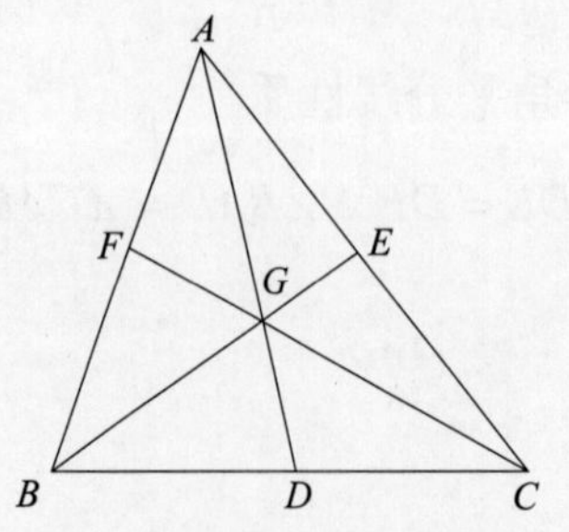

- Middle line of the triangle 三角形中位线性质

$$DE=\frac{1}{2}BC,\ \mathrm{DE//BC}$$

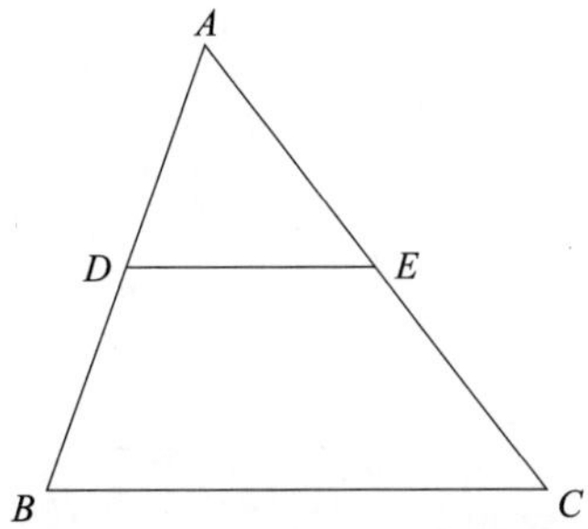

- Isosceles triangle: Same sides, same angles. Three lines in one.

Equilateral triangle: Same side length, same angles. Area $=\frac{\sqrt{3}}{4}a^2$. (a = side length)

The median from a right triangle: $BD=\frac{1}{2}AC$.

- 等腰三角形：两腰相等，两底角相等，三线合一。

等边三角形：边长相同，角度相同。面积 = $\frac{\sqrt{3}}{4}$边长的平方。（a = 边长）

直角三角形斜边中线等于斜边的一半：$BD=\frac{1}{2}AC$。

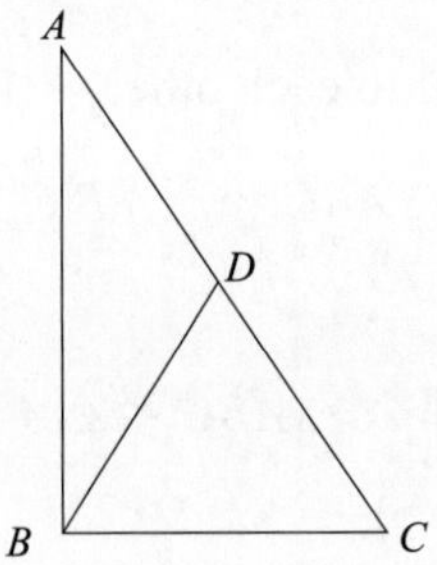

- Isosceles right triangle

The hypotenuse is $\sqrt{2}$ times of two right-angled sides, and both acute angles are 45°.

$$AB=BC=\frac{\sqrt{2}}{2}AC$$

$$\angle BAC=\angle BCA=45°$$

- 等腰直角三角形

斜边是两直角边的 $\sqrt{2}$ 倍，两锐角都是 45°。

$$AB = BC = \frac{\sqrt{2}}{2}AC$$

$$\angle BAC = \angle BCA = 45°$$

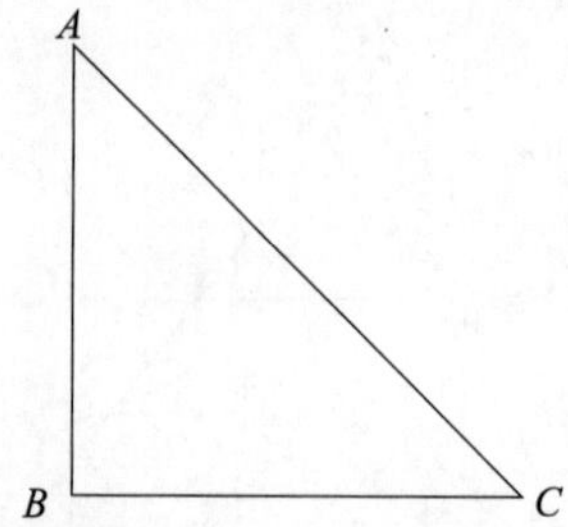

2. Similar Triangles 相似三角形

- Determination of Similar Triangles 相似三角形的判定

Principle 1 (SSS)　原理 1 (SSS)

Method 1: Proportionality of three sides

If $\frac{AB}{A'B'} = \frac{BC}{B'C'} = \frac{CA}{C'A'}$, then $\triangle ABC \backsim \triangle A'B'C'$.

方法一：三边对应成比例

如果 $\frac{AB}{A'B'} = \frac{BC}{B'C'} = \frac{CA}{C'A'}$，那么 $\triangle ABC \backsim \triangle A'B'C'$。

Principle 2 (AA)　原理 2 (AA)

Method 2: Two angles are equal to each other

If $\angle A = \angle A', \angle B = \angle B'$, then $\triangle ABC \backsim \triangle A'B'C'$.

方法二：两角对应相等

如果 $\angle A = \angle A'$，$\angle B = \angle B'$，那么 $\triangle ABC \backsim \triangle A'B'C'$。

Principle 3 (SAS)　原理 3 (SAS)

Method 3: Two sides correspond to be proportional and their angles correspond to be equal

If $\frac{AB}{A'B'} = \frac{BC}{B'C'}$, and $\angle B = \angle B$, then $\triangle ABC \backsim \triangle A'B'C'$.

方法三：两边对应成比例，它们的夹角对应相等

如果 $\frac{AB}{A'B'} = \frac{BC}{B'C'}$，且 $\angle B = \angle B$，那么 $\triangle ABC \backsim \triangle A'B'C'$。

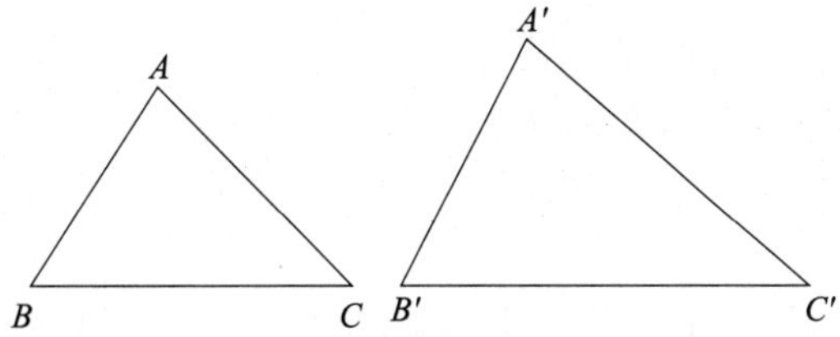

Conversely, if two triangles are similar, the proportionality of the above sides and the equivalence of the angles also hold true.

反之，如果两个三角形相似，以上边的比例关系和角的相等关系也成立。

- The properties of similar triangles 相似三角形的性质

(1) The ratio of corresponding sides of similar triangles is called scale factor. The ratio of perimeters of similar triangles is equal to the scale factor. The ratio of areas of similar triangles is equal to the square of the scale factor. The formulas are as follows:

If $\triangle ABC \backsim \triangle A'B'C'$, $\dfrac{AB}{A'B'}=\dfrac{BC}{B'C'}=\dfrac{CA}{C'A'}=k$, then

$$\frac{\text{perimeter of }\triangle ABC}{\text{perimeter of }\triangle A'B'C'}=k,\ \frac{\text{area of }\triangle ABC}{\text{area of }\triangle A'B'C'}=k^2$$

相似三角形对应边的比例称为相似比。如果两个三角形相似，那么三边对应成比例，并且周长比等于相似比，面积比等于相似比的平方。具体公式如下：

如果 $\triangle ABC \backsim \triangle A'B'C'$，且 $\dfrac{AB}{A'B'}=\dfrac{BC}{B'C'}=\dfrac{CA}{C'A'}=k$，那么

$$\frac{\triangle ABC\text{的周长}}{\triangle A'B'C'\text{的周长}}=k,\ \frac{\triangle ABC\text{的面积}}{\triangle A'B'C'\text{的面积}}=k^2$$

(2) Type A is similar to Type X A 型和 X 型相似

If $AB /\!/ A'B'$, then $\triangle ABC \backsim \triangle A'B'C'$.

如果 $AB /\!/ A'B'$，那么 $\triangle ABC$ 相似于 $\triangle A'B'C'$。

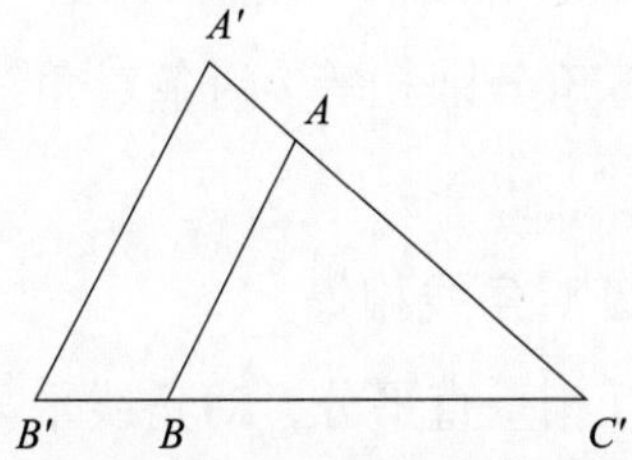

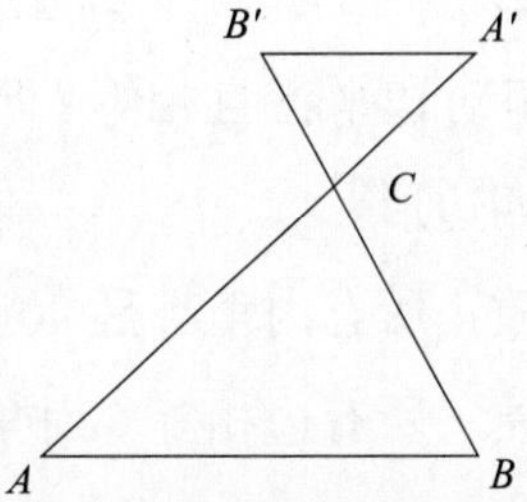

(3) Projective. Theorem in similar triangles 相似三角形中的射影定理

If $\angle DFE=\angle FGE=90°$, then

$$\triangle DFE \backsim \triangle FGE$$

$$FG^2 = DG \times GE,\ FG \times DE = DF \times FE$$

$$DF^2 = DE \times DG,\ EF^2 = DE \times EG$$

如果 $\angle DFE$ 和 $\angle FGE$ 都是直角，那么 $\triangle DEF$ 和 $\triangle FGE$ 相似，且有

$$FG^2 = DG \times GE, FG \times DE = DF \times FE$$

$$DF^2 = DE \times DG, EF^2 = DE \times EG$$

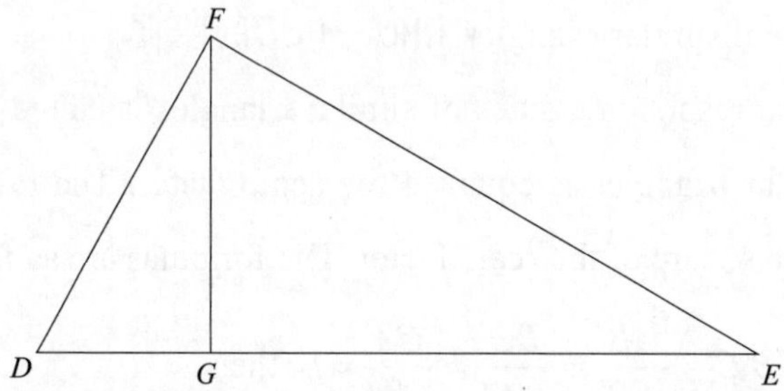

3. Parallelograms 平行四边形

- Properties of parallelograms: opposite sides parallel and equal; opposite angles are equal; diagonals bisect each other.
- Determinations of parallelograms: a quadrilateral that has two pairs of opposite sides parallel / two pairs of opposite sides equal / one pair of opposite sides parallel and equal / two pairs of opposite angles equal / two diagonals bisect each other, is a parallelogram.
- Properties of rectangles: all angles = 90°; diagonals are equal.
- Properties of rhombus: all sides are equal; diagonals bisect each other at 90°; diagonals bisect angles.
- 平行四边形的性质：对边平行且相等，对角相等，对角线互相平分。
- 平行四边形的判定：一个四边形，如果它的两组对边分别平行 / 两组对边分别相等 / 一组对边平行且相等 / 两组对角分别相等 / 两条对角线互相平分，那么它是平行四边形。
- 矩形的性质：所有内角都是 90°, 两条对角线相等。
- 菱形的性质：所有边相等，对角线互相垂直平分，对角线平分内角。

4. Polygons 多边形

- A polygon with all its sides and all its angles equal is called a regular polygon.

- Number of symmetry axis of a n-side regular polygon: n.
- order of rotational symmetry of a n-side regular polygon: n.
- Sum of interior angles of a n-side polygon $=(n-2)\times 180°$.
- Sum of exterior angles of a n-side polygon=360°.
- Number of diagonals of a n-side polygon $=\dfrac{n(n-3)}{2}$.
- Interior angle of a n-side regular polygon $=\dfrac{(n-2)\times 180°}{n}$.
- Exterior angle of a n-side regular polygon $=\dfrac{360°}{n}$.
- Center angle of a n-side regular polygon $=\dfrac{360°}{n}$.
- 所有边和角都相等的多边形叫作正多边形。
- 边数为 n 的正多边形有 n 条对称轴。
- 边数为 n 的正多边形是 n 阶旋转对称图形。
- 边数为 n 的多边形内角和为 $(n-2)\times 180°$。
- 边数为 n 的多边形外角和等于 360°。
- 边数为 n 的多边形的对角线个数为 $\dfrac{n(n-3)}{2}$。
- 边数为 n 的正多边形的内角大小为 $\dfrac{(n-2)\times 180°}{n}$。
- 边数为 n 的正多边形的外角大小为 $\dfrac{360°}{n}$。
- 边数为 n 的正多边形的中心角大小为 $\dfrac{360°}{n}$。

5. Circles 圆

- Circumference of a circle $=\pi d=2\pi r$.
- Area of a circle $=\pi r^2=\dfrac{1}{4}\pi d^2$.
- In the same circle or congruent circles, if two central angles / inscribed angles / arcs / chords / chord lengths of two chords are equal, then the others are all equal.
- The degree of the inscribed angle is equal to half of the degree of the central angle the arc it is facing.
- For a sector, we have

$$\frac{\text{arc length}}{\text{circumference}}=\frac{\text{central angle of the arc}}{360°}=\frac{\text{area of sector}}{\text{area of circle}}$$

- 圆周长 $=\pi d=2\pi r$。
- 圆面积 $=\pi r^2=\frac{1}{4}\pi d^2$。
- 在同一个圆或全等的两个圆中，如果两个圆心角 / 两个圆周角 / 两条弧 / 两条弦 / 两段弦长相等，满足其中一条，其它的也都成立。
- 圆周角等于它所对的弧对应的圆心角的一半。
- 在扇形中，弧长在整圆周长中所占的比，等于它所对的圆心角在 360° 中所占的比，也等于它的面积在整圆面积中所占的比。

6. Analytical Geometry 解析几何

- The distance between two points: $d=\sqrt{(x_1-x_2)^2+(y_1-y_2)^2}$.
- 两点间距离：$d=\sqrt{(x_1-x_2)^2+(y_1-y_2)^2}$。

7. Solid Geometry 立体几何

- For a prism, $V-E+F=2$: V stands for the number of vertices, E stands for the number of edges, F stands for the number of faces.
- surface area of cuboid $=6a^2$.
- Volume of cuboid $=a^3$.
- Surface area of a prism = 2 × area of cross-section + perimeter of cross-section × length.
- Volume of a prism = area of cross-section × length.
- Surface area of a cylinder $=2\pi rh+2\pi r^2=2\pi r(h+r)$, where r is the radius of the base circle and h is the height of the cylinder.
- Volume of a cylinder $=\pi r^2h$.
- Volume of a pyramid $=\frac{1}{3}\times$ base area × perpendicular height.
- Surface area of a cone $=\pi rl+\pi r^2=\pi r(l+r)$, where r is the radius of the base circle and l is the slant height.
- Volume of a cone $=\frac{1}{3}\pi r^2h$.
- Surface area of a sphere $=4\pi r^2$, where r is the radius.
- Volume of a sphere $=\frac{4}{3}\pi r^3$.
- 在棱柱中，顶点数－棱数＋面数 =2。

- 正方体表面积：$6a^2$。
- 正方体体积：a^3。
- 棱柱表面积 = 2 倍横截面面积 + 横截面周长 × 棱柱的长。
- 棱柱体积 = 横截面面积 × 棱柱的长。
- 圆柱表面积 $= 2\pi rh + 2\pi r^2 = 2\pi r(h+r)$，其中 r 是底面圆的半径，h 是圆柱的高。
- 圆柱体积 $= \pi r^2 h$。
- 正四棱锥体积 $= \frac{1}{3} \times$ 底面积 × 垂高。
- 圆锥表面积 $= \pi rl + \pi r^2 = \pi r(l+r)$，其中 r 是底面圆的半径，l 是斜高。
- 圆锥体积 $= \frac{1}{3}\pi r^2 h$。
- 球表面积 $= 4\pi r^2$，其中 r 是球的半径。
- 球体积 $= \frac{4}{3}\pi r^3$ 。

Counting and Probability 计数与概率

1. Sum Rule (Additional Principle) 加法原理

If an event E_1 can happen in n_1 ways, event E_2 can happen in n_2 ways,$\cdots$, event E_k can happen in n_k ways, and if any event E_1, E_2, $\cdots$, E_k happens, the job is done, then the total ways to do the job is $n_1 + n_2 + \cdots + n_k$.

加法原理：如果一件事情有 k 种不同的做法，其中做法 E_1 有 n_1 种方式完成，做法 E_2 有 n_2 种方式完成，……，做法 E_k 有 n_k 种方式完成，并且做法 E_1，E_2，$\cdots$，E_k 都有可能出现，那么这件事情的总完成方式共有 $n_1 + n_2 + \cdots + n_k$ 种。

2. Product Rule (Multiplication Principle) 乘法原理

When a task consistes of k separate parts, if the first part can be done in n_1 ways, the second part can be done in n_2 ways, and so on through the kth part, which can be done in n_k ways, then total number of possible results for completing the task is given by the product: $n_1 \times n_2 \times \cdots \times n_k$.

如果一件事情分成 k 个步骤完成，其中第一步有 n_1 种方式完成，第二步有 n_2 种方式完成，直到第 k 步，有 n_k 种方式完成，那么完成整件事情的方法数是：$n_1 \times n_2 \times \cdots \times n_k$。

3. Properties of probability 概率的主要性质

(1) The probability of an event is between 0 and 1.

(2) The probability of an impossible event is 0.

(3) The probability of a certain event is 1.

(1)任意事件的概率都在 0 到 1 之间。

(2)不可能事件的概率为 0。

(3)必然事件的概率为 1。

4. Basic probability formula 概率的基本公式

$$\text{Probability}=\frac{\text{number of ways th at a certain outcome can occur}}{\text{total number of possible outcomes}}$$

The probability that an event will occur is equal to one minus the probability that it will not occur.

$$\text{概率}=\frac{\text{特定事件的数量}}{\text{所有事件的数量}}$$

一件事情发生的概率等于 1 减去它不发生的概率。

5. Permutation and Combination 排列组合

The factorial of a positive integer n $n!=n\times(n-1)\times\cdots\times 3\times 2\times 1$

正整数 n 的阶乘 $n!=n\times(n-1)\times\cdots\times 3\times 2\times 1$

There are a total of P_n^r number of ways to permute r different ones from among n different objects.

n 个不同物体中选出 r 个排序,共有 P_n^r 种方法:

$$\mathrm{P}_n^r=\frac{n!}{(n-r)!}(1\leqslant r\leqslant n)$$

There are a total of C_n^r number of ways to choose r different ones out from n different objects.

n 个物体中选出 r 个,共有 C_n^r 种方法:

$$\mathrm{C}_n^r=\frac{n!}{r!(n-r)!}(0\leqslant r\leqslant n)$$

Permute n objects, among which p objects are of the first type, q objects are of the second type, r objects are of the third type and so on, then there are $\frac{\mathrm{P}_n^n}{p!\times q!\times r!\times\cdots}=\frac{n!}{p!\times q!\times r!\times\cdots}$ number of different permutations.

n 个物体排列,其中 p 个物体是一类,q 个物体是另一类,r 个物体是第三类,以此类推，共有 $\dfrac{P_n^n}{p!\times q!\times r!\times\cdots}=\dfrac{n!}{p!\times q!\times r!\times\cdots}$ 种排列方式

$$\frac{P_n^n}{p!\times q!\times r!\times\cdots}=\frac{n!}{p!\times q!\times r!\times\cdots}(p+q+r+\cdots=n)$$

The total number of ways to choose some (1 to n) objects from among n different ones is:

n 个物体中选取 1 到 n 个物体，所有方法数之和为

$$C_n^0+C_n^1+\cdots+C_n^n=2^n$$

6. Sets and Venn Diagrams 集合和韦恩图

The formula for the union of two sets: The number of elements in the union of Set A and Set B = The number of elements in set A + The number of elements in set B − The number of elements in the intersection of set A and set B.

两个集合的并集公式：集合 A 和集合 B 的并集中的元素数 = 集合 A 中的元素数 + 集合 B 中的元素数 − 集合 A 和集合 B 交集中的元素数。

$$n(A\cup B)=n(A)+n(B)-n(A\cap B)$$

The formula for the union of three sets: The number of elements in the union of Sets A, B and C = The number of elements in set A + The number of elements in set B + The number of elements in set C − The number of elements in the intersection of set A and set B − The number of elements in the intersection of set A and set C − The number of elements in the intersection of set B and set C + The number of elements in the intersection of set A, set B and set C.

三个集合的并集公式：集合 A、集合 B 和集合 C 的并集的元素数 = 集合 A 中的元素数 + 集合 B 中的元素数 + 集合 C 中的元素数 − 集合 A、B 的交集的元素数 − 集合 A、C 的交集的元素数 − 集合 B、C 交集的元素数 + 集合 A、B 和 C 共同交集的元素数。

$$n(A\cup B\cup C)=n(A)+n(B)+n(C)-n(A\cap B)-n(B\cap C)-n(A\cap C)+n(A\cap B\cap C)$$

Number Theory 数论

1. Odd and Even 奇偶性，Odd 奇数，Even 偶数

even ± even = even	偶数 ± 偶数 = 偶数
odd ± odd = even	奇数 ± 奇数 = 偶数
even ± odd = odd	偶数 ± 奇数 = 奇数
even × even = even	偶数 × 偶数 = 偶数

odd × odd=odd	奇数×奇数＝奇数
even × odd = even	偶数×奇数＝偶数
even ÷ even =even/odd/fraction	偶数÷偶数＝偶数/奇数/分数
even ÷ odd =even/fraction	偶数÷奇数＝偶数/分数
odd ÷ odd = odd/fraction	奇数÷奇数＝奇数/分数
even ≠ odd	奇数≠偶数

- The sum of two consecutive integers is odd: $n+(n+1)=2n+1$.
- The product of two consecutive integers is even: $n\times(n+1)=2k$ for some integer k.
- Any two consecutive integers have opposite parity.
- If the product of n positive integers is even, at least one of these n positive integers is even.
- If the product of n positive integers is odd, all of these n positive integers are odd.
- If the number of odd integers is even, then the sum of them is even.
- If the number of odd integers is odd, then the sum of them is odd.
- 两个连续整数的和是奇数 $n+(n+1)=2n+1$。
- 两个连续整数的乘积是偶数，即存在整数 k 使得 $n\times(n+1)=2k$。
- 两个连续整数有相反的奇偶性（一奇一偶）。
- 如果 n 个正整数的乘积是偶数，那么这 n 个正整数中至少有一个是偶数。
- 如果 n 个正整数的乘积是奇数，那么这 n 个正整数都是奇数。
- 偶数个奇数的和是偶数。
- 奇数个奇数的和是奇数。

2. Divisibility rule for 2, 4, 8, 16, and 2^n 2 的幂的整除性

A number is divisible by 2 if the last digit of the number is divisible by 2;

A number is divisible by 4 if the last 2 digits of the number is divisible by 4;

A number is divisible by 8 if the last 3 digits of the number is divisible by 8;

A number is divisible by 16 if the last 4 digits of the number is divisible by 16;

A number is divisible by 2^n if the last n digits of the number is divisible by 2^n.

如果一个数的最后一位能被 2 整除，那么这个数就能被 2 整除；

如果一个数的最后两位能被 4 整除，那么这个数就能被 4 整除；

如果一个数的最后三位能被 8 整除，那么这个数就能被 8 整除；

如果一个数的最后四位能被 16 整除，那么这个数就能被 16 整除；

如果一个数的最后 n 位能被 2^n 整除，那么这个数就能被 2^n 整除。

3. Divisibility rule for 3 and 9 3 和 9 的整除性

A number is divisible by 3 if the sum of the digits of the number is divisible by 3;

A number is divisible by 9 if the sum of the digits of the number is divisible by 9.

如果一个数的数位之和能被 3 整除，那么这个数就能被 3 整除；

如果一个数的数位之和能被 9 整除，那么这个数就能被 9 整除。

4. Divisibility rule for 5, 25, 125, and 5^n 5 的幂的整除性

A number is divisible by 5 if the last digit of the number is divisible by 5;

A number is divisible by 25 if the last 2 digits of the number is divisible by 25;

A number is divisible by 125 if the last 3 digits of the number is divisible by 125;

A number is divisible by 5^n if the last n digits of the number is divisible by 5^n .

如果一个数的最后一位能被 5 整除，那么这个数就能被 5 整除；

如果一个数的最后两位能被 25 整除，那么这个数就能被 25 整除；

如果一个数的最后三位能被 125 整除，那么这个数就能被 125 整除；

如果一个数的最后 n 位能被 5^n 整除，那么这个数就能被 5^n 整除。

5. Divisibility rule for 11 11 的整除性

If the difference between the sum of all digits in odd digit and the sum of all digits in even digit can be divised by 11, then the number is divisible by 11.

如果一个数的奇数数位之和与偶数数位之和的差能被 11 整除，那么这个数就能被 11 整除。

6. Divisibility rule for a composite number with at least 2 different prime factors 有大于等于 2 种不同质因数的合数的整除性

Divisibility rule for 6, 12, 14, 15, 18, 24 6，12，14，15，18，24 的整除性

A number is divisible by 6 if the number is divisible by both 2 and 3;

A number is divisible by 12 if the number is divisible by both 3 and 4;

A number is divisible by 14 if the number is divisible by both 2 and 7;

A number is divisible by 15 if the number is divisible by both 3 and 5;

A number is divisible by 18 if the number is divisible by both 2 and 9;

A number is divisible by 24 if the number is divisible by both 3 and 8.

一个数如果能被 2 和 3 整除就能被 6 整除;

一个数如果能被 3 和 4 整除就能被 12 整除;

一个数如果能被 2 和 7 整除就能被 14 整除;

一个数如果能被 3 和 5 整除就能被 15 整除;

一个数如果能被 2 和 9 整除就能被 18 整除;

一个数如果能被 3 和 8 整除就能被 24 整除。

To tell if a number is divisible by a composite divisor, it suffices to find the prime factorization of the divisor and tell if the number can be divided by the powers of the prime factors respectively. If it can be divided by all the powers, then it is divisible by the composite divisor.

判断一个数能否被一个至少有两种质因数的合数整除，只要把这个合数分解质因数后拆分成每个质因数的幂再分别判断，如果都可以整除，那么这个数就能被该合数整除。

7. Prime Factorization 分解质因数

- If a number can be written as $a = p_1^{b_1} p_2^{b_2} \dots p_n^{b_n}$, then the number of positive factor of is $(b_1+1)(b_2+1)\cdots(b_n+1)$; the sum of its factor is $\left(p_1^0+p_1^1+\cdots+p_1^{b_1}\right)\left(p_2^0+p_2^1+\cdots+p_2^{b_2}\right)\cdots\left(p_n^0+p_n^1+\cdots+p_n^{b_n}\right)$.
- Define $\lfloor x \rfloor$ is the largest integer less than or equal to x. Then the exponent of the power of prime number p in the prime factorization of $n!$ is $\left\lfloor \frac{n}{p} \right\rfloor + \left\lfloor \frac{n}{p^2} \right\rfloor + \left\lfloor \frac{n}{p^3} \right\rfloor + \cdots$
- 如果一个数为 $a = p_1^{b_1} p_2^{b_2} \cdots p_n^{b_n}$，这个数的正因子的数量等于 $(b_1+1)(b_2+1)\cdots(b_n+1)$；这些正因子的和为 $\left(p_1^0+p_1^1+\cdots+p_1^{b_1}\right)\left(p_2^0+p_2^1+\cdots+p_2^{b_2}\right)\cdots\left(p_n^0+p_n^1+\cdots+p_n^{b_n}\right)$.
- 定义 $\lfloor x \rfloor$ 是小于等于 x 的最大整数，则 $n!$ 的质因数分解中质因数 p 的次数为 $\left\lfloor \frac{n}{p} \right\rfloor + \left\lfloor \frac{n}{p^2} \right\rfloor + \left\lfloor \frac{n}{p^3} \right\rfloor + \cdots$.

8. GCF and LCM 最大公因数和最小公倍数

If $(a,b)=d$, and n is a positive integer, then $(na,nb) = nd$.

$$(a,1)=1;\ (a,a)=a;\ (a,b)=(b,a);\ (a,b)=(b,a-b)$$

$$\mathrm{GCF}(a,b)\times \mathrm{LCM}(a,b)=a\times b$$

如果 a 和 b 的最大公约数是 d，n 是一个正整数，那么 na 和 nb 的最大公约数为 nd。

最大公约数还满足以下性质：

$$(a,1)=1;\ (a,a)=a;\ (a,b)=(b,a);\ (a,b)=(b,a-b)$$

$$\mathrm{GCF}(a,b)\times \mathrm{LCM}(a,b)=a\times b$$